# Sensehacking

# Sensehacking

*How to Use the Power of Your Senses for Happier, Healthier Living*

CHARLES SPENCE

VIKING
*an imprint of*
PENGUIN BOOKS

VIKING

UK | USA | Canada | Ireland | Australia
India | New Zealand | South Africa

Viking is part of the Penguin Random House group of companies whose addresses can be found at global.penguinrandomhouse.com

First published 2021

001

Copyright © Charles Spence, 2021

The moral right of the author has been asserted

Set in 12/14.75 pt Bembo Book MT Std
Typeset by Jouve (UK), Milton Keynes
Printed and bound in Great Britain by Clays Ltd, Elcograf S.p.A.

The authorized representative in the EEA is Penguin Random House Ireland, Morrison Chambers, 32 Nassau Street, Dublin D02 YH68

A CIP catalogue record for this book is available from the British Library

HARDBACK ISBN: 978–0–241–36113–9
TRADE PAPERBACK ISBN: 978–0–241–36114–6

www.greenpenguin.co.uk

Penguin Random House is committed to a sustainable future for our business, our readers and our planet. This book is made from Forest Stewardship Council® certified paper.

To Babis, for keeping faith

# Contents

| | |
|---|---|
| List of Illustrations | xiii |
| 1. Introduction | 1 |
| Interior design for the multisensory mind | 5 |
| 'Scared you!' Why you like what you do and dislike what you don't | 6 |
| Sensory crosstalk | 11 |
| The merging of the senses | 13 |
| The science of sensehacking | 16 |
| 2. Home | 18 |
| Designing for the multisensory mind | 21 |
| 'Sensory living' | 24 |
| What was it with avocado and chocolate bathrooms? | 26 |
| The colour of emotion | 28 |
| Why do we like our homes to be as warm as Africa? | 29 |
| A kitchen for the senses | 30 |
| Is Baker-Miller pink really an appetite suppressant? | 32 |
| Can a nicely decorated table really make the food taste better? | 34 |
| Do you really want a silent kitchen? | 36 |
| Home alone | 37 |
| Getting into hot water | 38 |
| 3 Garden | 42 |
| The nature effect | 45 |
| A room with a view: decomposing the nature effect | 50 |

Ranking nature's benefits — 54
Natural timing — 56
Santandercito, my garden retreat — 59
Sensehacking nature for well-being — 62

4. Bedroom — 65

The shorter you sleep, the shorter you live — 66
Nodding off — 68
Dazzled by the light — 69
Sweet dreams? — 72
Sleeping soundly — 72
Sleep on it — 74
Sleep tight — 75
Why not pop a pot plant on your bedside table? — 76
Are you an owl or a lark? — 77
The first night effect — 78
Sleep deprivation — 80
The scent of sleep — 83
Sensehacking our dreams — 84
Rise and shine: is it time to wake up and smell the bacon? — 85
Sleep inertia — 86
Can dawn light wake you up? — 89
Are we really more sleep deprived than ever? — 90

5. Commuting — 92

What's real and what's not? — 93
Vroom, vroom: just how important is the sound of the engine? — 95
Can you hear the quality? — 97
'Make it snuggle in the palm' — 98
What has techno got to do with RTAs? — 99

Distracted by technology: capturing the attention of
the distracted driver ... 100
'It can wait' ... 102
Asleep at the wheel ... 104
Driving the nature effect ... 106
A risky solution ... 111
'Scared sick': why exactly do we get car sick? ... 112
The road ahead ... 115

6. Workplace ... 116

Sensory imbalance in the workplace ... 118
Anyone for lean design? ... 118
Talk about a sick building ... 119
Is the air conditioning sexist? ... 121
Who hasn't felt tired at work? ... 123
Sensehacking creativity ... 126
Open-plan offices ... 128
Sensehacking the open-plan workspace ... 129
Bringing nature to the workplace ... 132
'What's wrong with plastic trees?' ... 133
All the beauty of nature on your desktop? ... 135
What is the connection between creativity
and commensality? ... 139

7. Shopping ... 142

Leading the customer by the nose ... 144
Smelling colour ... 148
Moving to the beat ... 150
Subliminal seduction ... 154
Atmospherics ... 156
Smells like teen spirit ... 157
Keeping a cool head ... 160

'Touch me' ... 161
Tactile contamination ... 163
Can multisensory marketing really deliver a superadditive sales boost? ... 164
Time to turn the lights up, and the music down? ... 167
Tasting the future ... 169
Multisensory shopping online ... 170
On the future of online marketing ... 171
Shop 'til you drop ... 172

8. Healthcare ... 174

Why hospitals are starting to look like high-end hotels ... 176
Tastes healthy ... 180
Looks healthy: the art and illusion of healthcare ... 183
Healthy hearing ... 188
Music therapy ... 192
Healing hands ... 193
Scent-sory healing ... 195
Multisensory medicine: processing fluency and the dangers of sensory overload ... 197

9. Exercise and Sport ... 199

Is it better to exercise in nature or indoors? ... 201
Distraction for action ... 203
Mood music: moving to the beat ... 205
Why do tennis players grunt? ... 209
Listen to the sound of the crowd ... 211
The scent of victory, the taste of success ... 212
The power of clothing ... 215
Seeing red ... 217
Working-out with the senses ... 220

| | | |
|---|---|---|
| 10. | Dating | 222 |
| | Arousal | 223 |
| | 'The look of love' | 227 |
| | Scent of a woman: sexy smells | 230 |
| | 'The Lynx effect' | 232 |
| | Why is the lady always in red? | 235 |
| | Killer heels | 237 |
| | Swipe right: tips for online dating | 239 |
| | Are oysters really the food of love? | 240 |
| | The voice of desire: is beauty really in the ear of the beholder? | 242 |
| | Welcome to the smell dating agency | 243 |
| | Multisensory magic: I love you with all my senses | 244 |
| 11. | Coming to Our Senses | 247 |
| | Sensory deprivation | 248 |
| | Are you suffering from sensory overload? | 249 |
| | Primal pleasures: the nature effect | 251 |
| | Welcome to the sensory marketing explosion | 252 |
| | Sensism: a mindful approach to the senses | 253 |
| | Do you have the sensory balance right? | 255 |
| | Social isolation in the pandemic era | 257 |
| | Sensehacking new sensations | 259 |
| | Coming to our senses | 263 |
| | Appendix: Simple Sensehacks | 265 |
| | Acknowledgements | 267 |
| | Notes | 269 |
| | Bibliography | 301 |
| | Index | 353 |

# Illustrations

p. 8: An example of anthropomorphism in product design. (© Karim, Lützenkirchen, Khedr and Khalil)

p. 10: Some examples of brands that have incorporated a smile into their logo. (Amazon, Thomson/Tui, Hasbro)

p. 24: The first show kitchen completed by John Nash in 1818.

p. 31: The NozNoz. (NozNoz)

p. 59: My home in the Colombian cloud forest.

p. 121: Emma, the office worker of the future.

p. 135: Amazon's flagship offices in downtown Seattle. (Joe Mabel)

p. 143: The 7UP logo. (Jetijones)

p. 155: An example of subliminal symbolic messaging.

p. 166: Sub-additivity in the marketplace.

p. 187: How a mirror box can be used to trick a phantom limb patient into seeing their missing limb. (US Navy photo by Mass Communication Specialist Seaman Joseph A. Boomhower)

p. 229: The Venus of Willendorf. (MatthiasKabel)

p. 240: *Lunch with Oysters and Wine* by Frans van Mieris.

p. 248: Many 'high value' detainees were subjected to sensory deprivation at Guantánamo Bay. (US Navy)

p. 254: Arthur Dove's *Fog Horns* (1929). (yigruzeltil)

p. 257: A series of shocking experiments conducted in the 1950s by Harry Harlow on baby monkeys.

p. 258: The Hug Shirt. (Cute Circuit)

p. 259: Human cyborg, or should that be 'eyeborg', Neil Harbisson. (Dan Wilton)

p. 261: Cyborg Nest's North Sense. (Eugene Dyakov)

# 1. Introduction

From the moment we are born until the last breath we take, sensation is fundamental to our existence. Everything we perceive, experience and know comes to us through our senses. As Francis Galton, Charles Darwin's half-cousin, noted in 1883, 'The only information that reaches us concerning outward events appears to pass through the avenue of our senses; and the more perceptive the senses are of difference, the larger is the field upon which our judgement and intelligence can act.'[1]

Yet, paradoxically, what many of us complain about is sensory overload. We are all tired of being bombarded by too much noise, too much information, and too many distractions.[2] Just think about how much multitasking you do these days. According to a 2015 report by the business consultancy Accenture, 87 per cent of us use multiple media devices at the same time.[3] And the problem is only getting worse, with the world apparently getting faster, faster.[4] Look closely, though, and it soon becomes apparent that it is mostly our higher rational senses, namely, our hearing and vision, that can communicate large amounts of information and which are easily targeted by technology, that are being overstimulated. It is much harder to find anyone complaining about having to deal with too many smells,* too much touch, or else an overabundance of taste. It is, in other words, all a matter of getting the sensory balance right.[5]

North American researcher Tiffany Field, from the Touch Research Institute and Miller School of Medicine in Miami, has

---

* At least, not since the plethora of scent strips in magazines has reduced.

been arguing for years that most of us are 'touch hungry', and that this may be leading to a range of negative outcomes for our health and well-being. The skin, by far the largest organ we sense with, accounts for something like 16–18 per cent of our body mass.[6] In recent years, researchers have discovered that the hairy skin – that is, all the skin on your body except for the palms of your hands and the soles of your feet – is packed with sensory receptors that like – no, that *need* – to be gently stroked.[7] Such warm interpersonal touch conveys a wide range of benefits for our social, cognitive and emotional well-being; interpersonal touch between couples is so powerful that it can even help to relieve physical pain. The stress-buffering social support provided by frequent hugging also reduces one's susceptibility to upper respiratory tract infection and illness.[8] And if you don't have anyone to cuddle, don't worry; there are now professional huggers who, for a fee, will give you one. So, neglect your biggest sense at your peril. The Covid-19 pandemic set many of us thinking how we might deliver the touch that we all so urgently need from afar. We will return in the final chapter to the question of how technology may be used to deliver this interpersonal touch at a distance.

It has been estimated that more than a billion people will be over sixty years of age by the year 2025. Many older people complain about a lack of stimulation because no one wants to touch them since they have become, they say, physically unattractive.

Aging takes its toll on all our senses, though the decline starts at different ages for each of them. Fortunately, we have hearing aids and glasses to ameliorate the deterioration of sight and sound but, crucially, there is as yet nothing that can be done to restore our more emotional senses of touch, smell and taste. Consequently, the elderly are in very real danger of suffering 'sensory underload'.[9]

If you ask which of their senses they would miss the most,

most people immediately say their sight. However, the data from quality-of-life indicators and suicide rates indicate that those who have lost their sense of smell are actually often much worse off. After all, many of those losing vision later in life can still picture their loved ones when hearing their voices. This can help to soften the blow of sensory loss, at least for a while. But when we lose the ability to smell, there really is nothing left of that sense. Few amongst us inhabit a world of remembered smells that is anything like as rich as in the mentally imagined world that we can conjure before our mind's-eye.[10]

Over the last quarter of a century or so, I have been lucky enough to work with many of the world's largest companies translating the emerging science of the senses into implementable strategies to help promote health and well-being, not to mention profitability (obviously). This includes everyone from Johnson & Johnson to Unilever, from Asahi to the VF Corporation, and from Dulux to Durex.[11] In the pages that follow, I want to share with you what I have learned.

I have spent many years working with paediatricians to highlight the importance of ensuring a balanced mix of multisensory stimulation for the optimal social, emotional and cognitive development of babies.[12] I have helped car manufacturers to sensehack driving, and so make our roads safer.[13] (I will tell you more about what I found out in the chapter on Commuting.) I also work closely with many of the world's largest beauty, fragrance, home and personal care and sexual health companies trying to figure out how multisensory attraction really works.\* (Don't worry, I'll pass on some of my top tips in the Dating chapter.)[14] I have also consulted extensively for many high-street

---

\* It is worth noting the prediction made in 2016 that the global cosmetics market would be worth a not inconsiderable $675 billion by 2020. (It is, of course, likely that the Covid-19 pandemic will have reduced that figure in practice.)

brands and mall owners across the globe, devising new ways to entice you to spend money at their 'emporia of the senses'. (We will learn more about this in the Shopping chapter.) These, then, are just a few of the secrets of the senses that I will be sharing with you.

So what exactly is sensehacking? It can be defined as using the power of the senses, and sensory stimulation, to help improve our social, cognitive and emotional well-being. It is only by recognizing the unique capacities of every one of our senses, and by acknowledging the predictable ways in which they interact to guide our feelings and behaviours, that we can hope to hack our own sensory experience most effectively. By so doing, we can all start to improve the quality of life of those we care about, starting with ourselves. No matter whether you want to feel more relaxed or more alert; no matter whether you want to be more productive or less stressed at work; no matter whether you want to sleep better, figure out how to look your best, or else get the most out of your workout at the gym; the science of sensehacking is here to help you to achieve your goals and aspirations.

The chapters in this book are organized around the key activities of daily life, and a number of the most frequently encountered everyday environments in which we are likely to find ourselves, at least if your life is anything like mine. We start at your front door, in the aptly named Home chapter. We will take a look inside the living room, the kitchen and the bathroom in order to see how each of our senses can be hacked to make our homes more hospitable and more habitable, not to mention easier to sell. We then move on to look at the benefits of nature in the Garden chapter. Next, in the Bedroom chapter, we take a look at how to sensehack our sleep by playing with perception, in literally every sense. This is a particularly important issue, since so many of us currently complain about problems getting enough sleep, and an especially worrying one, given the stats concerning how bad for our health and well-being missing out on sleep is. Having covered the home environment, I move on to work, sensehacking Commuting and the multisensory design of the Workplace. Finally, leisure is covered in chapters on Shopping, Healthcare, Exercise and Sport, and Dating. In each case, we will take a look at some of the most effective sense hacks that have been demonstrated to make us spend more, exercise harder, look better, and recover from disease or injury more quickly.

In the concluding chapter I will recap some of the key issues and insights around the themes of sensory overload, sensory underload, sensory balance, multisensory congruency and the technology-mediated sensorium in which we find ourselves increasingly often. As a Professor of Experimental Psychology at Oxford University, my insights and recommendations are based firmly on hard science from peer-reviewed academic research, as opposed to the often unsubstantiated claims from various lifestyle gurus, feng shui 'experts', interior designers and trends futurologists so often found in this genre, so you are in safe hands.

## Interior design for the multisensory mind

Getting the multisensory atmosphere right in the places where we live and work really matters, especially once you realize that those of us living in an urban environment, which is now the majority of the world's population, spend 95 per cent of our time indoors. As we will see in several of the subsequent chapters, this results in a sensory imbalance that is having a negative effect on our well-being. Spending so much of our time indoors not only can all too easily restrict our access to natural light but also means that many of us are being exposed to more airborne pollutants in poorly ventilated office buildings than is good for us. As we will see in the Workplace chapter, there is mounting evidence that spending so much time indoors, as is currently the case for many of us, may be leading to problems such as sick building syndrome and seasonal affective disorder, the latter affecting something like 6 per cent of the population in the UK who suffer from insufficient natural light in the dark winter months. Given the move towards open-plan offices in recent years, we will also look at what you can do to make sure that the sensory attributes of the working environment help, rather than hinder, your productivity and creative output. Many of the world's largest paint, lighting and fragrance companies have long been interested in trying to design multisensory interiors that allow us to achieve our goals, whatever they might be.[15]

Over the years, I have consulted for gyms and a range of other sporting organizations, helping people to get the most out of their exercise and fitness regimes by engaging the power of the senses to help motivate, energize and distract. We will take a closer look at this in the Exercise and Sport chapter. Gaining a competitive advantage, be it on the sports field or in one's love life, requires us all to make the most of whatever our senses have to offer. A number of innovative individuals and organizations are already making themselves healthier, wealthier and wiser by using the latest insights from the world of sensehacking. What is more, they are doing all this without the side effects that are usually associated with pharmacological interventions – what some refer to as 'cosmetic neuroscience'.[16] So what are you waiting for?

## 'Scared you!' Why you like what you do and dislike what you don't

Familiarity breeds liking. Did you know that simply by exposing you to something, I can increase your liking for it? This is known as the 'mere exposure effect', and it works no matter whether we are aware of having been exposed or not.[17] Mere exposure presumably explains why it is that some of us like to eat chillies while others prefer to listen to the Red Hot Chili Peppers. It also helps to explain why it is that newborn babies turn their heads toward the aromas of the foods that their mothers consumed during pregnancy. Have you heard of 'foetal soap syndrome'? Don't worry, it's not as bad as it sounds. The term was introduced in the 1980s by doctors who noticed that some newborn babies preferred the voices of popular soap stars at the time (think Kylie Minogue and Jason Donovan in *Neighbours* if you're old enough) to the dulcet tones of their own mothers. It turned out that the babies had not only been tasting what their mothers had been eating but also listening to what they had been listening to.[18] Such insights and observations are leading some to wonder just how early in life the senses can be hacked.

While the majority of our responses to the multisensory stimuli that surround us are learned, it is always important to bear in mind that we evolved within a particular niche. This means that those stimuli once important for our survival seem to have retained a special status. For instance, it turns out that spiders and snakes can capture our attention and make us feel anxious, even from a very young age.[19] And, as we will see in the Home chapter, the latest evidence suggests that we also tend to set our central-heating thermostat to imitate the climatic conditions in the Ethiopian highlands where we evolved many millennia ago.

The more of your senses that are stimulated by nature, the better. Furthermore, being attentive to all that the environment has to offer our senses, rather than passing mindlessly through your surroundings, can magnify this benefit still further. Time and again in the chapters to come we will see how capturing some aspect of nature, often mediated by technology, can have a positive impact, no matter whether we happen to be at work, exercising, out shopping or at play. In order to know how best to hack our senses, no matter whether it is with the aim of enhancing our own well-being, or else perhaps nudging us to buy more whenever we go out shopping, it surely has to make sense first to try to understand the evolutionary niche in which our senses developed. Or, to adapt the famous line from geneticist and evolutionist Theodosius Dobzhansky, nothing in psychology makes sense except in the light of evolution.[20] For instance, in the Garden chapter, we will see how important those small green spaces outside the home are in providing us with a dose of the nature effect.

We also have a hard-wired, or evolved, preference for smiles over frowns, no matter whether we see them on other people's faces or on clock faces. There are stimuli that we appear to be evolutionarily prepared either to like or to loathe. Successful design and marketing often makes use of such subtle, and not so subtle, triggers to nudge us consumers by means of the latest in sensory marketing, as we will see in the Shopping chapter.[21]

Why is 8.20 a downer? Which clock face is smiling at you? Which watch looks sad? This is just one example of anthropomorphism in product design.

Have you, for instance, ever noticed, or wondered why it is, that analogue watches in ads nearly always show the time as ten minutes past ten? In fact, according to the results of one analysis, ninety-seven of the one hundred bestselling men's dress watches on Amazon.com, showed this time. The position of the hands at 10.10 makes it seem like it is smiling, and when researchers in Germany assessed this experimentally, they found that people prefer those watches that 'smile' at them.[22]

Those who market timepieces long ago realized that simply by setting the time to 10.10 on an analogue watch they could make us feel more favourably about what they were offering us.* Such a simple sense hack really shouldn't influence our choice – after all, the precision of a watch's timekeeping is not signalled by the time it shows – but the evidence suggests that it most certainly does. What is more, having recognized the importance of such evolutionary triggers, one can then design everything from product packaging to computers and cars, safe in the knowledge that they will be appreciated that little bit more by consumers because they play on specific evolutionary affordances. For instance, designers have shown that arranging the USB slots on the front of your computer, or the visual design of cars, so that they look, anthropomorphically, like they are smiling can also help to create a better, more appealing impression too. Once you know what to look out for, it will suddenly become apparent just how many companies are already hacking your senses. For instance, look closely and I bet you will be amazed to see how many brands are smiling at us these days, everyone from Amazon to Argos.[23]

Intriguingly, however, watches did not always smile. The

---

* And, for those of you out there wondering why the time is 9.42 a.m. in most iPhone commercials, the most popular suggestion is that this was the time (Pacific Standard) when Steve Jobs first introduced the iPhone at a MacWorld conference back in 2007.

Some examples of brands that have incorporated a smile into their logo.

majority of adverts from the 1920s and 1930s displayed watches set to 8.20 (frowning). Hence, while making one's product smile might seem obvious today (at least to the intuitive marketers and designers of this world), such evolutionarily inspired solutions were often stumbled upon by accident. And while some have, rightly in my view, wanted to question the ethics of sensory marketing and consumer neuroscience, in this strange twenty-first century, when global pandemics and lockdowns are apt to devastate the global economy without warning, I would argue that never has there been an era when sensehacking was more important. Sensehacking is what we need to remain well and balanced.[24]

Indeed, during the deepest pre-Covid-19 recession, the Great Depression that followed the Wall Street Crash of 1929, 'consumer engineering', or 'humaneering', was regarded by some as a key element in terms of helping to reinvigorate the world economy. This novel approach to consumer psychology was all about getting the subtle sensory cues in product design right.[25] In a way, then, consumer engineering can be seen as last century's precursor to today's sensehacking. However, in order to hack them most effectively we first need to recognize that our senses do not work independently but talk to each other all the time. And it is only by understanding the key rules governing how the senses interact to deliver our multisensory perception of the world around us, and all the objects within, that we can really begin to optimize the multisensory cues and environments to help deliver the outcomes we all desire.

## Sensory crosstalk

When I first started teaching in Oxford almost a quarter of a century ago, there was one professor studying vision and another studying hearing. Even though these two individuals worked together closely (both physically and metaphorically), they had fallen out and not spoken to one another for years. The truly surprising thing to me, though, was that neither of them appeared concerned at this lack of communication. They simply did not realize what they were missing out on. Implicit in their attitude was the traditional view of perception, whereby the senses were considered as entirely separate systems. This is, after all, just how they appear to us on the outside: we have eyes to see, ears to hear, a nose to smell, a tongue with which to taste, and skin to feel the world around us.

But how much do your senses interact? The answer to this question is critical, both in terms of determining how we experience the world around us and how that experience makes us feel. The science shows that our senses connect far more than we ever imagined. In practice, this means it is possible to change what we hear simply by altering what we see, while manipulating the way something sounds can affect what it feels like. And simply by adding the right scent, it is possible to bias our impression of whatever is before our eyes, as the wily marketers know only too well. Any one of us can also use the tricks, or sense hacks, associated with multisensory perception, to our own advantage.

In the chapters that follow, we will come across many such surprising examples of crosstalk between the senses. For instance, in the Dating chapter, we will see that how attractive, not to mention how youthful, you *look* to others is partly determined by your choice of fragrance (i.e. how you *smell*). We will also see how the agreeable sensation of a loving caress can be ruined by a bad smell while being enhanced by a good one.[26] In the Home chapter, we will discover how you can make your washing feel softer and look whiter, simply by adding the right scent. We will also see how our perception of the bitterness of coffee turns out to be influenced as much by the harshness of the noise made by the coffee machine as by your choice of beans and roast. And, in the Commuting chapter, we will see how your perception of the quality of a car is subtly, not to say subliminally, influenced by the psychoacoustically engineered sound of solidity made by its doors when you close them, hence priming notions of security. If you drive an upmarket model, then the chances are that the throaty roar of the engine that you so admire has actually been synthesized to sound a certain way too. All that before we get to the recommendations to add value to an older model using nothing more than a dash of 'new-car smell'.

While big business has been hacking our senses for decades,[27] there is nothing to stop the rest of us from using sensehacking to help us to eat less without feeling hungry. Wouldn't you want to know if it turned out that you could use 'warm' lighting and/or paint colours to help reduce your heating costs? We can all get more out of life by hacking our senses. This book is going to show you *how* to do it, and *why* it works. Sensehacking provides one of the most effective means of helping us to eat less, live longer and enjoy ourselves more along the way. We can all use music, soundscapes, scents and colours to become more productive, to relax and sleep better and to improve perception when necessary.[28]

Sensehacking is built on the growing awareness of just how connected our senses are, and how fundamental the right balance of sensory stimulation is to our health, productivity and well-being. This is true no matter whether we happen to be at home, in the office, at the gym, out shopping or even receiving healthcare. Integrating the senses is fundamental and, what is more, may even help to improve the quality of life too. In fact, correctly balanced multisensory stimulation is already being used in some clinical settings to help alleviate pain and aid in the recovery of hospital patients. We will come across a number of remarkable examples of such sensehacking in the Healthcare chapter, such as how patients who listen to music not only require less pain relief but may even recover faster too.

## The merging of the senses

No creature that has more than one sense keeps them separate. Just think about it for a moment. It would be catastrophic if one sense pulled in one direction while another pulled in the opposite direction. There is simply no way of resolving such a conflict unless the senses communicate. Our perception and behaviour are controlled by the activity of many millions of multisensory neurons connecting the five main senses of sight, sound, smell, touch and taste. The key question is what rules the brain actually uses to combine the inputs from the different senses. For it is only by understanding how multisensory perception works that we will be ready to start hacking our senses efficiently. In the Home and Dating chapters we will come across many examples of how sights, sounds, smells and even feelings combine to deliver the extraordinary, as well as the mostly mundane, multisensory experiences of everyday life. But what are the rules governing multisensory perception? Well, the good news is that there are only three key rules that you need to know about for now.

### 1) Sensory dominance: Hearing what you see

Very often, one of our senses will dominate the others in terms of dictating what we perceive. For instance, just think about how the actor's voice always seems to come from their lips on the cinema screen. This despite the fact that the voices actually emerge from loudspeakers hidden away somewhere else in the auditorium. In this case, our brain uses the evidence before our eyes to infer where the sounds belong. The 'ventriloquism effect', as it is known, has been used in stage shows, not to mention by mystics, for millennia. This makes sense inasmuch as our eyes typically do a much better job of telling us where something is than our ears. During human development, our brains learn to rely on the most dependable, or accurate, sense, thus helping us to deal with the 'blooming, buzzing confusion'* that greets us all at birth.[29] While some researchers have wanted to explain vision's dominance in purely mathematical (specifically Bayesian) terms, there remains some intriguing evidence suggesting that there is a more deep-seated reliance on, or attention to, whatever is before our eyes, that the mathematicians have not yet been able to explain.[30] This is where the anthropologists, historians, artists and possibly even sociologists can help us to contextualize the hierarchy of the senses that we currently find ourselves with, and enable us to probe

---

* According to William James (1890), one of the godfathers of experimental psychology, and someone who seemingly had a quotable line for every occasion: 'The baby, assailed by eyes, ears, nose, skin, and entrails at once, feels it all as one great blooming, buzzing confusion.'

whether it is the right one for us personally, not to mention for the societies in which we live.[31]

Imagine watching a person's lips articulating the syllable 'ga' while at the same time hearing them utter the syllable 'ba'. What would you perceive? Most people hear the speaker saying 'da'.[32] You can try this illusion, known as the McGurk effect, for yourself. There are plenty of great examples online. Very often our brains combine the various sensory inputs automatically, and without letting us know quite what is going on. Sometimes, even when you know exactly what is happening, and that your senses are tricking you (as in the McGurk effect), it is still impossible to stop yourself from hearing something different when the lip movements you see change their tune. Exactly the same thing happens when red dye is added to a white wine and the experts suddenly start smelling what they take to be red, or rosé, wine aromas.[33] The question of which sense dominates when they get conflicting messages is one that we will return to throughout the book.

## 2) Superadditivity: or when 1 + 1 = 3

Have you ever noticed just how much easier it can be to hear what someone is saying at a noisy cocktail party if you just put your glasses on?* Many researchers believe that this provides an everyday example of superadditivity. This is the idea that individually weak sensory inputs may sometimes combine to give rise to a multisensory experience that is much richer than would be predicted simply by the sum of the individual parts. As we

---

* According to the research, it is possible to deliver something like a 15dB increase in speech intelligibility in noise simply by looking at the associated lip movements.

will see in the Shopping chapter, marketers have become very excited by the possibility that they can enhance sales simply by getting the combination of in-store sounds, scents and colours right.

### 3) Sensory incongruence

Do you remember the last time you watched a badly dubbed foreign movie, say, or else a satellite broadcast where the voice became desynchronized from the lips? The visual image may be crystal clear, the voice quality second to none, but if the senses do not quite match up temporally, the experience can all too easily be ruined. Sub-additivity, the third of the key rules governing multisensory perception, is often the result when incongruent sensory impressions are combined. In fact, the outcome can all too easily end up being worse than would have been expected from the best of the individual inputs. Incongruent combinations of sensory stimuli are typically hard to process, meaning that they are negatively valenced; in other words, we do not like them much.* When the owner of the store or shopping mall starts playing the sounds of the jungle, or forest, over the loudspeakers, is that congruent? This just one of the questions that we will return to in the chapters that follow.

## The science of sensehacking

By avoiding sensory overload, sensory imbalance and sensory conflict, every one of us has the sensory tools at our disposal

---

* There is a great short video created by the Future of Storytelling (FoST) Institute in New York in 2016, should you wish to find out more about the rules of multisensory perception: https://futureofstorytelling.org/video/charles-spence-sensploration.

to help us live healthier, happier and more fulfilling lives by building on the emerging scientific understanding of the synergy between the senses. That requires us to recognize the cultural construction of the senses, acknowledge the inherently multisensory nature of mind and also be sensitive to the individual differences in the sensory worlds we all both inhabit and desire. This is the multisensory science of sensehacking. Now let's see how it is done.

## 2. Home

Let's start at the entrance. Have you ever wondered about the slightly strange scent that greets your nostrils as soon as you open your front door on returning from holiday? In fact, this smell is always there, it is just that we all adapt to it since we are continually exposed to it. It is only when we return after a long trip away that we suddenly realize what our own home actually smells like. Everyone's home has its own distinctive building odour (or BO, for short), as you've probably noticed, and your own abode is, of course, no different. Because we spend much of our time there, it becomes so familiar that we do not necessarily perceive it as a visitor would. All too often, we miss what is, quite literally in this case, right under our noses.[1]

But what effect does constant exposure to that smell have on your household? After all, the ambient scents that surround us can affect our mood and well-being, not to mention how alert or relaxed we feel.[2] And just because you are not aware of it certainly does not mean that you are immune to scent's influence upon you. Indeed, smells that we don't notice can sometimes affect us more than those that we do. Worryingly, there is also evidence to suggest that the airborne mould and other unidentified low-level odorants in people's homes may be responsible, at least in part, for the debilitating condition known as sick building syndrome, whereby the building in which you work and/or live can actually make you ill.[3]

Baudelaire once described the smell of a room as 'the soul of the apartment',[4] and I am sure that you have heard the advice to brew some coffee, bake some bread or a cake and/or put out some fresh flowers when prospective buyers view your home. The smell of vanilla is apparently particularly popular with North American real-estate brokers. Now, while it is actually surprisingly hard to find any concrete evidence showing that such a strategy works, this has not stopped others from going even further when it comes to scenting homes they are trying to sell. Indeed, according to one press report, we should all forget about the smell of coffee and fresh bread, because the perfect smell to sell a new home actually consists of a mixture of white tea and fig. Once again, though, no evidence was provided to back up this particular claim. At the same time, however, the smell of cigarette smoke and pets – and here we are talking cats and dogs, not goldfish – can be particularly off-putting, potentially reducing sale prices by as much as 10 per cent according to the CEO of one Florida estate agency.[5]

In 2018 the lucky owners of one high-end apartment commissioned synaesthetic* scent-designer Dawn Goldsworthy to create a bespoke scent especially for the $29 million new condo that they had just bought in Sunny Isles Beach, Miami. The idea was for this fragrance to be dispersed via the apartment's air-conditioning system, thus giving their property a truly unique olfactory identity. While such a strategy is obviously beyond the reach of all but the super-rich, there is certainly nothing to stop the rest of us from scenting our homes more frugally, be it with flowers, pot pourri or perhaps one of those battery-powered scent-dispensers. Scented candles are another option, although

---

* Synaesthesia is the unusual condition in which one sensory input consistently and automatically gives rise to another, idiosyncratic 'sensory concurrent', as when people see colours when looking at black and white letters or hearing music. Synaesthesia may be somewhat more common amongst creative individuals than in the rest of the population.

some commentators have started to raise concerns about their air-polluting effects, so burn them at your peril.

Many florists now offer 'well-being bouquets' – flower arrangements that not only look great but also give off scents that can exert a positive influence on our mood and our health.[6] So, no matter whether you're trying to sell your home or just giving it a spring clean, thinking more carefully about the scent that pervades the space might be a good idea. It certainly makes sense when one considers all the evidence showing that aromatherapy can influence our mood, our alertness and even our sense of well-being. Similarly, the presence of a pleasant fragrance such as warm bread, roast coffee or citrus can even increase our prosocial behaviour, be it measured in terms of helping others when they drop something or else tidying up after ourselves when we finish eating.[7]

The effects of ambient aroma appear to be more psychological than pharmacological in nature, meaning that they result from associative learning.[8] So, for example, it is likely to be the psychological associations we have that explain why the smell of heliotropin is so calming to so many of us. The scent of this South American flower is one of the key volatiles in Johnson's baby powder. The smell probably triggers reassuring memories and associations of early childhood, even if we cannot quite place it when coming across it out of context as adults. Meanwhile, some older people take a dislike to the scent of certain flowers because it reminds them of what lies ahead for us all: death and funerals. More generally, our response to ambient scent may well be determined by whether we happen to like the smell or not, whether we believe it to be artificial or natural, and how intense, or overpowering, we find it, although the published studies in this field are often statistically underpowered, meaning that it is hard to draw any firm conclusions from them.[9]

## Designing for the multisensory mind

Years ago, the modernist Swiss architect Le Corbusier made the intriguing claim that architectural forms 'work physiologically upon our senses'.[10] The latest findings from the emerging field of research lying at the intersection of cognitive neuroscience and architecture have started to provide some robust empirical support for such a claim.[11] For instance, the researchers in one study had their participants perform a social stress test in virtual reality. The psychologists' favourite stress test, known as the Trier Social Stress Test (or TSST for short), after the city in Germany where it was developed, involves having people give a short speech in front of a panel of listeners who show no emotion whatsoever. The panellists are usually real people, but in this case virtual characters were used instead. The researchers documented an increase in the level of salivary cortisol, which is a physiological marker of stress, in those who undertook the stress test in an enclosed virtual environment as compared with those who performed the task in a virtual room that gave the appearance of being more open. In this case, the suggestion was that the more open room provided a better opportunity for escape.[12]

For what it is worth, entering a home that smells bad can trigger a desire to escape too. People rate rooms as brighter, clearer and fresher if there is a low level of ambient fragrance present – even if they are unaware of the scent. What is more, rooms are experienced as larger when a pleasant scent is present. It has even been suggested that an ugly interior can be made to look tasteful by adding the right fragrance – though I must admit I have my doubts about that.[13]

The preference that we all show for open rather than enclosed rooms has been linked to *habitat* and *prospect-refuge* theories, first put forward by evolutionary psychologists almost half a century ago. Their suggestion was that we tend to prefer those landscapes and built environments that contain features that would once have aided our species' survival. According to refuge theory, for example, we like those environmental features that make us feel safer. Perhaps this explains why we have, for millennia, been bringing nature into our homes in the form of potted plants (because a large plant offers a potential hiding place). The practice goes back at least to the Egyptians in the third century BC, and houseplants were also found amongst the ruins at Pompeii.[14] That said, and as we will see later, indoor plants may also play a useful role in helping to purify the air, and can calm us down as a result of the nature effect that was mentioned earlier.

We are all drawn more toward round forms than to angular ones, and so one of the fundamental factors influencing how welcoming and attractive the rooms in our home appear is determined both by their shape and by their furnishings.[15] Ingrid Fetell Lee, former design director at IDEO in New York, argues that as you move through your home, angular objects exert an unconscious effect on your emotions, even if they're not directly in your path. As she notes, angular shapes 'may look chic and sophisticated, but they inhibit our playful impulses. Round shapes do just the opposite. A circular or elliptical coffee table changes a living room from a space for sedate, restrained interaction to a lively center for conversation and impromptu games.' One even finds echoes of this in the Feng Shui notion that we should fill our homes with plants having rounded leaves rather than spiky/pointy ones – think cheese plant rather than palm or, worse still, a cactus.[16]

The benefits of curvilinear over rectilinear form extend to the kitchen table too. Surprising as it might seem, those sitting around a round table are more likely to come to agreement than those seated at a square or rectangular one. The suggestion here is that a circular seating arrangement primes a need to belong whereas an angular one primes a need to be unique.[17] And, should you want to have people sitting a little closer together, why not put on some music? For, according to a project sponsored by Apple and Sonos (companies with a vested interest in promoting our music consumption), having it playing at home results in us sitting 12 per cent closer to one another than when there is nothing playing (though whether that is just so we can all hear each other better isn't entirely clear).[18]

Given how many arguments are thought to start while families are seated at the dining table, thinking about table shape constitutes one simple step towards making the home situation a little less fraught.* To put this problem into context, according to the results of a survey of 2,000 parents commissioned by Tex-Mex food brand Old El Paso that was reported in 2017, the average family argues twice a night during the course of their evening meal. A further 62 per cent of respondents admitted that they argued at dinner every night of the week, though I imagine it will probably take more than investing in a round dining table to improve the situation in such cases.[19]

The height of the ceiling affects our thinking style too. Rooms with high ceilings tend to be rated as more appealing while also leading to, or priming, something of a loftier, or freer, style of thinking. In particular, we tend to engage in more relational thinking (i.e. about how things relate to each other) rather than item-specific, or 'confined', thinking.[20] The apparent increase in ceiling height that results from painting a ceiling

---

\* Of course, after nearly every home-cooked meal there is washing-up to contend with. For their 2001 multisensory campaign, Procter & Gamble, the company whose brands include Fairy washing-up liquid, commissioned research highlighting that one of the most stressful aspects of home life is deciding whose turn it is to wash the dishes. Their helpful solution to this perennial problem was to launch a range of new 'thicker', colourful washing-up liquids scented with aromatherapy oils in order to help calm what can be a stressful situation.

The first show kitchen, completed by John Nash in 1818 at the Brighton Pavilion for the Prince Regent, later George IV. The renowned French chef Marie-Antoine (Antonin) Carême initially worked below these imposing metal 'palms'. Carême was perhaps most famous for his formidable culinary architectural creations, arguably influenced by the architecture of the space he was working in. Japanese writer Jun'ichirō Tanizaki writes eloquently on the surprising link between Japanese architecture and the culinary arts in his book *In praise of shadows*.

white, rather than a darker shade, also helps to explain our preference for the former when it comes to interior decor.[21] One can only wonder what culinary flights of fancy must have been fostered in those working in the world's first show kitchen, at the Prince Regent's Royal Pavilion in Brighton. In this case, the ceiling was seven metres high and supported by columns resembling palm trees.

### 'Sensory living'

An advert appearing in *The New York Times* in 2019 promoting a new residential development by renowned architect Odile Decq promised residents 'Sensory living'. The newspaper's readers were warmly welcomed to an address, and hopefully a new apartment, that would 'provoke their senses', the ad continued. It is certainly striking how much attention legendary designers such as Ilse Crawford pay to the senses in their work. In her book, *Interior designing for all five senses*, North American interior designer Catherine Bailly Dunne emphasizes the need to appeal to, to tantalize even, each and every one of the senses by designing rooms that 'not only look good, smell good, feel good, sound good, and even taste good, but they have that indefinable "something" that makes you feel at home right away'. However, while the rest of us often talk about the 'feel' of our home, it is surprising how little effort we actually go to in order to ensure that we get it right across all of the sensory touch-points.[22]

Interior designers often stress the importance of texture contrast. We all intuitively associate emotions with different textures and materials. Even if we don't actually touch the walls or surfaces, incorporating different materials in our homes can still make a tactile impression on us. A common recommendation from sensehacking designers and architects is to use a range of natural textures that you feel compelled to touch.[23] For instance, I always like to have a conker or rough piece of bark on my desk, something natural to ground me to reality, contrasting with all those smooth clinical surfaces in my office.* Another common suggestion is to cover smooth chairs and sofas with textured throws.

At the same time, however, it is important to note that our perception of the feel of materials is also influenced by the ambient scent that may pervade the space they occupy. My team in Oxford, for instance, reported some years ago that fabric swatches can be made to feel softer simply by adding the right

---

* You can, I suppose, think of this as a pocket-sized version of tree-hugging.

scent. In our case it was the smell of lemon or lavender, as compared to an unpleasant synthetic animal odour, that made the fabrics appear softer. All the more reason, then, to ensure that your freshly laundered towels smell nice. It really can make all the difference to how they feel.[24] In 2019 a pair of German researchers even reported that playing soft music can influence people's ratings of the softness of materials.[25]

When entering the home, though, it is the colour scheme that people are likely to notice first. And talking of (terrible) colour schemes . . .

## *What was it with avocado and chocolate bathrooms?*

Readers of a certain age might recall the almost obligatory avocado and/or chocolate-coloured bathroom suites of the 1970s. What were people thinking? This is just one memorable example of how the most desirable colours for our homes are often determined more by trends or fashion than by any functional claims for the physiological effects that colour may have on us. Whenever I see such a colour scheme in a retro-styled or unmodernized home today, I find it hard to imagine why people once flocked to install them.* Nowadays, the big paint companies employ trend experts to come up with the most desirable new-season colour names to keep their offerings sounding fresh and to make their paints seem contemporary. We should perhaps all be thankful that they have yet to conjure up a winning new name for this particularly unsavoury pair of colours, one that makes us fall in love with them all over again.

Le Corbusier hinted at an almost moral angle to the use of

---

* I suppose evolutionary psychologists might argue that these colours replicate the green of nature and the brown of the earth!

colour, or rather to its absence, when writing in 1923, 'Demand bare walls in your bedroom, your living room and your dining room . . . Once you have put ripolin [whitewash] on your walls you will be master of yourself. And you will want to be precise, to be accurate, to think clearly.' Here Le Corbusier seems to have been using white for its purifying value, as a sort of architectural sanitizer, much as men in the eighteenth century would have worn a white shirt instead of washing their bodies.* I do, though, wonder what the great architect would have made of German painter Johannes Itten's claim, 'Colour is life; for a world without colours appears to us as dead.'[26]

People make more errors when proofreading in a white room than in one that is either red or blue, thus seemingly contradicting Le Corbusier's suggestions. Elsewhere, meanwhile, in a study of almost 1,000 workers from different countries, those who found themselves in a room with some colour tended to be in a better mood than those working in a room without any.[27]

A large Italian study published in 2018 investigated the impact of interior colour on psychological functioning in 443 students living in university halls of residence just outside Pisa. The corridors, kitchens, living areas, and part of the residents' rooms in six identical buildings were painted in six different colours and the students' impressions were collected a little over a year later. The most popular colour was blue,† followed by green, violet, orange, yellow and red. The preference for blue was much stronger amongst the male

---

* Nowadays, it is only the inside, if no longer the outside, of our fridges that still need to appear pristine white. That and, I guess, toilets.

† Though, given this study was conducted in Italy, one needs to remember that the Azzurri, as the national football team is fondly known, also play in a blue strip. This association was used in the 1980s to try and help explain why Italian men respond negatively to relaxing pills (i.e. meaning they become more aroused) if they are coloured blue. Everywhere else in the world, blue is seen as a relaxing colour.

students than amongst the females, while the sex difference was reversed for those living in the violet-coloured building. Those living in the blue building also reported that they found it easier to study than those inhabiting rooms of any other colour.[28]

## The colour of emotion

What colour paint should you put on the walls of your home? Your choice really matters, given that the colour of the rooms inside which we spend so much of our time has been shown to exert a significant impact on our mood and well-being, with effects being linked to the hue (red, blue, green etc.), saturation (i.e. chroma, the purity or vividness of colour) and brightness (referring to the dark or light nature) of the colour. One of my favourite illustrations of this comes from famous Italian movie director Michelangelo Antonioni, who once painted the canteen a bright red with the idea of getting his actors into the right frame of mind prior to filming some tense scenes. This simple change to the environment apparently proved so effective that within a few weeks fights had started breaking out amongst the canteen's regular diners. Consistent with such an anecdotal observation, laboratory-based research has documented a measurable increase in arousal, as measured by what is called the galvanic skin response (basically, a measure of how much you sweat), after no more than a minute's exposure to red light as compared to light of other colours.[29]

The colour of walls and/or lighting has been shown to bias our mood, emotion and arousal. It has even been suggested that our internal clocks run just that little bit faster under red lighting than under blue. Consider how for centuries performers waiting to go on stage used to wait in a 'green room' that was literally green. Indeed, it is just this ability of colour to sensehack our emotions that might lie behind room lighting's influence on the perceived taste and enjoyment of food and drink. So, for example, the authors of one study demonstrating an effect of the colour of the ambient lighting on the taste of wine suggested that 'if a colour induces a positive mood or emotion [. . .] then the same wine tasted in this positive mood is liked better than when in a negative mood'. The colour scheme in a room can, then, perhaps be considered in much the same way that we think about the background music. For there too, it has been shown that the more we like the music, the more we appear to like whatever we are tasting.[30] The colour of the ambient lighting also influences our perception of the brightness of the lighting as well as the ambient temperature, which brings us nicely on to the question of . . .

## Why do we like our homes to be as warm as Africa?

The mean temperature, not to mention the regularity and consistency of indoor heating, has increased dramatically since the nineteenth century, with night-time averages continuing to rise well into the modern era. For instance, the mean indoor dwelling winter temperature increased by as much as 1.3°C (2.3°F) per decade from 1978 through to 1996.* More than 85 per cent of homes in the US have air conditioning, further helping homeowners to control the ambient indoor temperature.[31] So the question that you have to ask yourself, given how many of us now have so much control over our indoor climates, is why we all seem to like our homes to be as warm as Africa? These days, most of us set our homes to a fairly uniform 17–23°C (63–73°F).

---

* Though here it is worth bearing in mind the possible influence on these figures of the oil crisis of the 1970s.

At least that was the result of a study in which researchers collected indoor climate data from the homes of thirty-seven 'citizen scientists' across the United States over the course of the year. These data were then compared with global terrestrial climate data looking for the closest match.

Remarkably, no matter whether the citizen scientist's home was in Hawaii or Alaska, the chilly north of Washington State or the humid south of the Florida Everglades, the average indoor temperature and humidity across the course of the year that matched most closely turned out to be the mild outdoor conditions of west central Kenya or Ethiopia, where human life is first thought to have evolved. What such results have been taken to suggest is that we set our own home environments to mimic that of our prehistoric ancestors. According to Mark Maslin, an academic from University College London, these results highlight the enduring effects of 5 million years of evolution in East Africa.[32]

## A kitchen for the senses

The kitchen is one of the spaces in the home that has perhaps changed its form and function more than any other over the last century and a half. In Victorian times, it would typically have been hidden well out of sight. However, concern about food smells percolating through the home was something that people worried about even then. In 1880 the architect J. J. Stevenson wrote that, 'unless the kitchen itself is ventilated so that all smells and vapours pass immediately away, they are sure to get into the house, greeting us with their sickly odour in the halls and passages, and finding their way to the topmost bedroom, notwithstanding all contrivances of swing doors and crooked passages'. Le Corbusier's solution to this problem at the Villa

Savoye, a modernist villa situated in Poissy, on the outskirts of Paris, that the Swiss architect designed together with his cousin Pierre Jeanneret, was to put the kitchen on the roof. The idea was that this would help avoid the home filling with cooking aromas at mealtimes.[33] Dispelling cooking odours was just the sort of luxury that people could start to worry about after the widespread introduction of flushing toilets (actually invented in 1596) in the middle of the nineteenth century.[34]

Nowadays, the kitchen/dining room has become the common living space where many of us spend much of our waking time while at home. Thus we are being increasingly often exposed to food cues in the home, both olfactory and visual. This may be inadvertently nudging us toward eating and drinking more than we otherwise might.[35] Do not despair, though, for the creative designers at NozNoz have come up with an inventive, if bizarre, solution to tackle this particular problem. According to the instructions, all you have to do is to insert

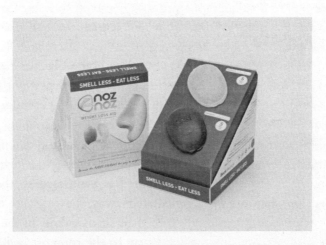

The NozNoz, one potentially effective means of sensehacking your sense of smell in order to avoid all of those tempting food aromas that so often surround us.

their patented rubber seals discreetly into your nostrils in order to block out all the homely food aromas that might otherwise be tempting you. *Voilà!* Problem solved. The results of a small preliminary study (not peer-reviewed as far as I can tell) suggested that wearing these inserts developed by the Rabin Medical Center in Israel could lead to a doubling in weight loss in dieters under fifty years of age. Over the three months of the trial, overweight individuals wearing the inserts apparently lost an average of 18lb (8.2kg; 7.7 per cent of body mass), as compared to 9.8lb (4.5kg) in the other group. I wait with bated breath to see whether this particular solution catches on – though something tells me that it perhaps will not. The good news, though, is this may not be the only sense hack that has been suggested to suppress our appetites.

### Is Baker-Miller pink really an appetite suppressant?

According to Kendall Jenner, painting your walls Baker-Miller pink acts as an appetite suppressant.* Jenner, the half-sister of Kim Kardashian, is currently one of the world's biggest Instagram stars (with more than 75 million followers last time I checked). When she painted her living-room walls in this colour after seeing an exhibition in which a pink Schauss Kitchen had been displayed it was big news, at least for the Instagram generation. However, before rushing out to stock up on this most unappetizing shade of bubblegum pink, it is worth noting, once again, that there is simply no support for this suggestion, intriguing though it undoubtedly sounds.

---

* This bubblegum shade is sometimes referred to as 'drunk tank pink', as it has been used in police holding cells for decades to calm rowdy detainees, based on some questionable research first published by Schauss in 1979.

It is not that the visual system does not hold certain wavelengths of light in special regard, for it most certainly does. For example, it turns out that the wavelength tuning of the cones in the retina is optimal for detecting the blood-oxygenation level in bare skin (regardless of skin colour), which provides a barometer of emotional arousal.[36] But important as it may be as a social signal, pink, and specifically Baker–Miller pink, is probably not the colour for your kitchen – so relax. Researchers have struggled to demonstrate any influence of this colour on people's behaviour in settings such as prison detention cells, and what is more, the effects that have been reported were only short-lasting anyway.[37]

Have you come across those colour-changing LED light bulbs? Twenty-four different colours or more in a single unit, but what to do with them all? Well, for those of you struggling in this regard, let me give you a little inspiration. According to my colleague Han-Seok Seo and others, coloured light *can* actually be used to help suppress our appetite. These researchers found that bathing men in blue light led to their eating significantly fewer omelettes and mini-pancakes for breakfast than those eating under regular white or yellow lighting. What is more, those who ate less under the blue lighting did not report being any less full after their diminished breakfast. One possible explanation for this intriguing result is that the blue lighting may have made the food look unappealing. This is the same claim that lies behind the blue-tinted dieting glasses sold by the Japanese company Yumetai.[38] For whatever reason, the appetite-suppressing effect of blue light was seen only in men. The women who took part in the study were not so easily swayed by such a cheap psychologist's 'trick of the light', perhaps because they tend to be rather more knowledgeable about what is in their food in the first place. However, before rushing out to buy a colour-changing light bulb for your own breakfast table, or perhaps sending off for a pair of those azure specs, it would, I think, be wise to wait for more robust studies to be published confirming just how long any appetite-suppressing effects of blue light last.

## Can a nicely decorated table really make the food taste better?

On one of my Colombian father-in-law Ricardo's final visits to England, my wife and I decided to take him out for a fancy molecular gastronomy meal in a very trendy, not to mention rather expensive, London restaurant. Unfortunately, what was meant to be a special occasion was ruined for my *suegro* (as they say in Spanish) by the paper napkins, not to mention the absence of a starched white linen tablecloth. As soon as Ricardo, a renowned architect back home, laid his eyes, and worse still his hands, on the rough feel of the napkin, it was all over; he simply could not enjoy the meal. The problem was that he just could not see past the table settings. And while this was frustrating at the time, given how much effort we had gone to in order to secure the table etc., the latest research has proved him right. A well-appointed table really can make the food taste better.

Coloured napkins, tablecloths and place settings can all exert a modest influence over a diner's emotional state in much the same way as changing the colour of the room.[39] Even the mere presence of a tablecloth can make a world of difference.* According to the results of a study published in 2020, diners rated a bowl of tomato soup as tasting significantly better when there was a tablecloth as compared to when it had been removed. The

---

* Sitwell (2020), Chapter 4, suggests that tablecloths first made their appearance in medieval England, being mentioned in a 1410 poem, 'London Lickpenny', where the author describes a Westminster establishment where, on approaching, 'a fayre clothe they begin to spread'.

enhancement in this case amounted to a not inconsiderable 10 per cent, and went together with a 50 per cent increase in how much was eaten – all this assessed in a reasonably naturalistic social dining situation where several people were eating together. The study also compared the influence of setting a nice table with that of varying the lighting intensity.[40] Traditionally, it was suggested that a darker, more intimate environment was better for dining, perhaps because it reminded us of that time long, long ago when, after a long day hunting bison or some such, the prehistoric family of the day could finally sit and relax in their cave in front of the fire and feel, for a few moments at least, safe. (Trust me, I am not making this stuff up.)[41] Interestingly in this regard, the nicely appointed table setting actually exerted a more pronounced effect on food ratings than did the lighting, which varied between romantic dark and bright light. Who says that table decorations do not matter?

So what, then, are some of my top tips for sensehacking the dining table to make that special meal really special? Classical music would seem like a good idea, given that it primes notions of quality and class. Soft lighting can't do any harm. My all-time favourite, though, has to be to make sure you get out the heaviest cutlery that you can find, as that will definitely help to ensure that your guests are impressed with whatever it is you have prepared, as I have described in my last book – *Gastrophysics: The new science of eating*. If you must have the TV on, try to watch something like *Downton Abbey* or perhaps an episode of *The Crown*. According to research commissioned by Aldi and conducted by Mindlab in 2013, watching *Downton* (and so also, presumably, anything in the same genre) results in people rating drinks such as wine, beer and brandy as tasting more elegant and refined. At least it did when compared with watching an episode of *Only Fools and Horses*, a sitcom set on a south London council estate. Once again, this is likely to be an example of sensation transference, much like the effects of pleasant music on fabric softness we encountered earlier.

## Do you really want a silent kitchen?

'The kitchen that sounds like a library' – this was the promise made in one print advertisement for AEG kitchens at the turn of the twenty-first century. Drawing our attention specifically to the quietness of the kitchen is unusual, and just like the silent car that features in the Commuting chapter, I am not so sure that it is something that any of us really want, at least not once we think about it. You would, I suspect, be taken aback if you knew just how much sensehacking goes into the design of our seemingly mundane white goods. Everything from the sound made by your fridge door when it closes through to the grinding, gurgling and steaming of your coffee machine has in all likelihood been engineered to sound 'just so'. Of course, when you realize that some kettles make more than 85dB of noise when boiling water (equivalent to heavy traffic, and the point at which long-term exposure starts to become damaging), it is no surprise that many of us complain that our electronic appliances make far too much noise and we would prefer them to be quieter.

Whether we realize it or not, sound often plays a functional role in our product experience. When engineers or designers do sometimes manage to remove it, thus making the washing machine, blender or vacuum cleaner silent, trouble often ensues. The problem is that it often feels like it is no longer doing a good job; for example, researchers have found that it is very difficult to convince people that a silent vacuum cleaner is picking up anything like as much dirt as a noisier model.[42] My team and I have been working for a number of years on modifying the sound of kitchen devices such as coffee machines. And surprising though it might sound, it turns out that it is possible to change people's perception of the harshness, or bitterness, of the coffee simply by changing the sharpness (i.e. the loudness and frequency profile) of the noise made by the machine. Getting the sound of the fridge door right when it closes is equally important at the high end of the kitchen appliance market. In fact, the design of fridge doors actually has more in common with the design of car doors than you might think. In both cases, having the right secure sound and feel is key. Perhaps this should not come as a surprise when one realizes that, at the upper end, your most desirable new designer kitchen can easily set you back more than even a fancy new car (as my wife will be pleased to confirm).[43]

## Home alone

In the 1930s, kitchen design evolved into creating an efficient working space, an especially important consideration after the domestic servants of former times had been dispensed with. Since the Second World War, though, there has been a shift from the kitchen as a secret place, never to be seen by visitors, as in the Victorian era, to one that is an integral part of our home's living space for so many of us. It has become the heart of the home, where we cook, work, hang out and do much of our entertaining.[44] Nowadays, it can sometimes feel as if the population can be divided into those who have chosen to invest in ever bigger and better show kitchens and those whose new apartments no longer have a kitchen at all. Indeed, confusion even reigns amongst house builders, as highlighted by the following quote from the Mayor of London's *Housing Space Standards* report from August 2006: 'there is continuing uncertainty over whether the kitchen space needs to be maintained, can be reduced (are meals only ever cooked in a microwave or are cookers and food preparation space still needed?), or (with households using more appliances) needs to be larger'.[45]

The trend towards removing kitchens from our homes certainly goes hand-in-hand with the rise in solo households. So many people living alone who would rather simply microwave a ready meal or have takeaway food delivered direct to their door. Indeed, one of the most profound changes in the home environment in recent decades has been the rapid and sustained growth in the number of single-person households. In Sweden, for example, 51.4 per cent of households comprise just one person (this is the highest figure in Europe). The figure in the UK is currently running at 31.1 per cent. The growth of solo living is leading to a worrying rise in loneliness, and at the same time more of us eating alone than ever before. Hence, one of the most important challenges in the years ahead concerns how we can sensehack commensality, most likely by connecting solo diners at home via digital technology. Figuring out how best to do this looks to me like one of the most intriguing challenges for the coming years.[46]

Having looked at the role of the senses in living and eating at home, all that it remains for me to do in this chapter is to take a quick look at what has to be one of the most multisensory rooms in the home, the bathroom.

## Getting into hot water

Only 4 per cent of the population apparently has enough time for a regular bath these days, with 76 per cent of us preferring the speed, cheapness and efficiency of a shower. According to one poll, a third of people in the UK bathe less often than once a week, with one in five admitting to never taking a bath at all.* To many, this preference for showering is perfectly understandable, especially as baths take longer. And bear in mind that those who choose to shower already waste six months of their lives standing under the showerhead. Six months – imagine![47]

Bathing is certainly one area where there are some profound individual, not to mention cultural, differences. While many Westerners like to start their day with a shower or a bath, in the Far East, for example in Japan, a soak is more usually enjoyed at the end of the day. This, as we will see in the Bedroom chapter, is perhaps not such a bad idea, given that exposure to hot water at the end of the day can help us get to sleep. North Americans love their power showers, finding the British attachment to wallowing in the tub 'unnatural'. And before any North American readers raise the issue, the traditional British fondness for bath time has nothing to do with the trickle that so many British shower heads offer, at least when compared with the pulverizing vigour of a standard North American attachment.

Here, and this is where I may find myself getting into some metaphorical hot water, I have to admit that I am one of those who are enamoured with the virtues of a regular long hot bath. I am definitely with the writer Sylvia Plath on this when she said, 'There must be quite a few things that a hot bath won't cure, but I don't know of many of them.' This, note, was something that one of our greatest leaders, Winston Churchill, was very partial to. He would take two deep, exceedingly hot baths a day, filled initially to 98°F (36.7°C), as checked by his thermometer-wielding assistant. Once Sir Winston was safely submerged, the temperature would then be increased to 104°F (40°C).[48] I suppose the more fundamental question, though, is why anyone should find taking

---

* This sounds rather like my grandfather on my mother's side, who took but one bath a year – the problem being that he didn't shower either.

a hot bath pleasurable in the first place?* Well, from a health perspective, research from both Loughborough University in the UK and Oregon in the States has shown that taking a hot bath not only helps to lower our blood pressure and reduce inflammation but also can help to burn calories. For instance, in one study, taking a one-hour bath of at least 104°F (40°C) was shown to burn off the same number of calories (140 to be precise) as walking fairly vigorously for half an hour. Meanwhile, going one step further (as it were) in terms of the heat stakes, research from Finland has shown that taking frequent saunas reduces the likelihood of heart attack or stroke in men.[49]

At the other extreme, starting the day with a bracingly cold shower is apparently also good for us. According to the results of a recent randomized trial, starting the day with a (hot-to-)cold shower† may actually benefit both our performance and our health. In particular, a 29 per cent reduction in self-reported sickness absence was documented in those adults who took a daily cold shower every day for at least a month.‡ To put this figure into perspective, regular exercise led to a 35 per cent reduction in sick leave. Silicon Valley tech entrepreneurs have apparently got in on the act too, with a growing number of them swearing by starting their day with a cold shower in order to induce what has been called 'positive stress'.[50] Just take Jack Dorsey, billionaire founder and CEO of Twitter, who reputedly

---

* The mind boggles as to what the evolutionary psychologists would come up with on this matter.
† Showering as warm as is comfortable for you for as long as you like, but then ending up with up to 90 seconds of cold water.
‡ In ancient Rome, bathing was based around the practice of moving through a series of heated rooms culminating in a cold plunge at the end. In modern times, the traditional ritual of the *frigidarium* has been kept in most saunas and spas around the world.

takes a couple of ice baths a day, sandwiched between saunas, to help improve his mental clarity.

Regardless of the health benefits, our bathing habits have been called into question in recent years, as commentators increasingly seem to want to nudge us all towards cutting our water consumption. (A bath will use something like eighty litres of water, an eight-minute shower uses around sixty-two.*) So far, that mainly consists of the users of bathrooms in older houses being encouraged to put a rubber brick in their toilet cisterns to reduce usage with every flush, but others are already thinking about whether they can hack our senses in order to nudge us all to use a little less water. Another possibility here may be to make the tap louder. In one study, this simple sense hack convinced people that the flow was higher, thus potentially helping remind them of all the water they were using.[51]

Have you noticed just how fragrant everything is, from shower gel to bath salts? Be aware, once again, that the olfactory component of our home and personal care products is very relevant, given that how silky soft and shiny we think our hair is after a wash (assuming that we have hair, that is), just like the softness of our towels, is again influenced by the scent.[52] Similarly, in one of my favourite studies, researchers demonstrated that it was the relaxing scent added to the moisturizer that was doing all the work in terms of removing facial wrinkles, albeit temporarily, not any of the active ingredients. So, now we are all nice and clean, and smelling of roses, it is time to put our shoes on and head out into the garden.

---

* Though it should be borne in mind here that some power showers may use as many as 136 litres (30 gallons), thus making the question of which form of bathing is more efficient a little less certain.

## 3.  Garden

I have a confession to make. I used to be a very naughty boy. From my earliest school days, I was shoplifting, letting off stink bombs and blowing things up in the chemistry lab, and receiving the appropriately named Mr Payne's plimsoll on the rear for my efforts when caught. I can still vividly remember the day I brought my own size 6 trainers to the headmaster's office, only for Mr Payne to bellow that those weren't shoes before bringing out his enormous size 13 footwear to gently caress my behind! Things went from bad to worse as the years went by until on one fateful day, aged thirteen or so, I was given a public flogging in front of all the teachers and prefects in school for 'throwing' concentrated nitric acid in another boy's face. In my defence, I only meant to brandish the bottle and give my nemesis (let's refer to him as A. S.) a scare while the chemistry teacher was out of the room. Unfortunately, however, I was holding the bottle by its glass stopper and it slipped from my grasp at the critical moment, along with its most corrosive contents, smashing onto the table and splashing my classmate in the process. I will never know whether it was this public humiliation, the last flogging in my school before the practice was outlawed in the UK, that changed my behaviour for the better. But whatever it was, my performance soon started to improve. Within a year or so I was starting to come top of the class rather than bottom and, well, the rest, as they say, is history.

In hindsight, there was another significant change that occurred at around this time: I joined the school's orienteering club. I was soon hooked. Before long I was spending most weekends and many an evening running through the woods, moors and forests of the North of England, map and compass in hand, searching out the red-and-white flags to stamp my card.* Suddenly I was being exposed to nature in all its richness on a near daily basis. Could it have been *this* change to my diet of multisensory stimulation that was responsible for the positive change in my behaviour and, ultimately, my school performance? It would never have crossed my mind to think so at the time, but a growing body of research now shows just how profoundly beneficial exposure to nature is for all of us, no matter who we are.

Even small doses of nature have been shown to improve our mood, our performance and our well-being.[1] Increase the amount of time spent in nature and the benefits rise in a seemingly dose-dependent manner. Perhaps, then, what really changed my behaviour, and I suppose my life, for the better all those years ago was not so much the pain and humiliation associated with a good hiding as all that exposure to nature.† As we saw in the opening chapter, the fundamental problem in the modern era is that we seem to have lost touch with our more sensual side. The increasing amounts of time that those living an urban existence, which is the majority of people these days, spend indoors mean that we are all in danger of losing touch with nature, and the multisensory benefits that it provides. As Marc Treib once wrote in an essay entitled 'Must landscape mean?' – 'Today might be a good time to once more examine the garden in relation to the senses.'[2]

For many of us, the garden is where much of our exposure to

---

* Orienteering is a kind of timed treasure hunt outdoors, just without the treasure.
† I will never know for sure as my attempts to run the controlled experiment by bringing back corporal punishment while I was Dean of Discipline at Somerville College, where I now teach, was (understandably) vetoed by the other fellows!

nature occurs.* Though, of course, local parks, woods and forests are also accessible to many of us. Intriguingly, the statistics suggest that while the number of people who have a garden is falling, those who do have one are spending more on gardening than ever before. In the UK it has been estimated that there will be 2.6 million homes without access to a private garden in 2020 (up from 2.16 million in 2010 and 1.6 million in 1995, representing a drop of around 2 per cent of households).[3] Meanwhile, in the US in 2018, gardeners spent a record $47.8 billion on lawn and garden retail sales, with a record average household spend of $503, up nearly $100 on the previous year's figure.[4]

In this chapter, I want to review the evidence concerning the benefits of the nature effect. I have already mentioned the positive effect of bringing plants into the home and later, in the Healthcare chapter, we will look into the healing gardens that have been a distinctive feature of certain hospitals for centuries. For now, though, to give a sense of the benefit of using one's garden, just take the latest research from the University of Exeter Medical School and the Royal Horticultural Society. The study of almost 8,000 UK residents with gardens revealed that 71 per cent of those who made use of their garden reported good health, as compared to just 61 per cent of non-garden users. This difference, which was evidenced in terms of higher mental wellbeing and increased activity levels, was equivalent to that seen when comparing those living in high-income areas and those from poorer neighbourhoods. So spending time in the garden, no matter whether gardening or just relaxing, really is good for us.[5] Much of the interest in the nature effect has been triggered by the eminent North American sociobiologist E. O. Wilson's

---

* Even better if you have a house close to the water, or as English actress Andrea Riseborough once put it, 'There's something really simple and idyllic about living in a house very close to the water.'

influential suggestion that we humans are biophilic, meaning that we have a natural affinity for living things.

## The nature effect

For millennia, people have known intuitively that exposure to nature is beneficial. Two thousand years ago, Taoists in the Far East were already advocating in print the health benefits of gardening and greenhouses.[6] Similarly, the residents of ancient Rome also valued their contact with nature as an antidote to all the noise, congestion and other stresses of city life.[7] Frederick Law Olmsted, one of the designers of Central Park in New York City, wrote in 1865, 'It is a scientific fact, that the occasional contemplation of natural scenes of an impressive character . . . is favorable to the health and vigor of men.'[8] In the more recent past, any number of romantic poets, novelists, philosophers and artists have managed to forge successful careers by drawing our attention to the joys of communing with nature. In fact, the strapline from an article that appeared in Finnair's inflight magazine *Blue Wings* in December 2019 captured the idea when stating, 'True luxury today is connecting with nature and feeling that your senses work again.'

Nowadays, in Japan and Korea, there are plenty of researchers encouraging anyone who will listen to take a 'forest bath' – known as *shinrin-yoku*. The term refers to a mindful immersion in our surroundings, paying close attention to the sights, sounds, smells and even feel of nature.* Forest bathers exhibit lower stress levels as well as an enhanced immune response.[9] And that is certainly nothing to be sniffed at, though mindful inhaling

---

* And no, that doesn't necessarily mean tree-hugging; though that may be good for you too!

is certainly part of what you'll be advised to do should you find yourself forest bathing. It is interesting to note the stress that is placed on the olfactory benefits of nature in the Far East. The focus of forest bathing is really on the inhaling of volatile substances called phytoncides. These antimicrobial organic compounds, derived from trees, include such wood essential oils as α-pinene and limonene. This focus on the olfactory contrasts noticeably with the stress that tends to be placed on the sights and sounds of nature in so much Western research in this area.

Recovery from laboratory-induced stressors occurs more rapidly following short-term exposure to nature. The standard way to induce stress in the lab is to give people an exceedingly difficult task to perform while informing them that most other people solved it easily; that, or else make them watch a stressful film. Heart rate and skin conductivity, both physiological markers of stress, drop back to baseline levels more rapidly after such stressors when we are in nature as compared to when we find ourselves in the built environment.[10]

The beneficial effects of nature operate across the lifespan. That said, pretty much every one of us living in a modern industrialized society is likely to be suffering from what has been called a 'nature deficit'.[11] Getting children into nature every now and again, no matter whether it is a little urban gardening during the school day or Outward Bound courses for recalcitrant youths such as I once was, is remarkably beneficial. In fact, given the evidence, we should *all* be engaging more with nature than is currently the case. In terms of the elderly, one study by Ottoson and Grahn published in the journal *Landscape Research* in 2005 reported that spending an hour in the garden in the geriatric home resulted in the inhabitants finding it easier to concentrate than when they had just spent the same amount of time resting indoors.

So much for the controlled experiments, but what about people's behaviour out in the wild, as it were – that is, in the real world? In one study, more than a million alerts were sent out to 20,000 people in the UK via their iPhones at various times of day over a six-month period. The phone's satellite GPS was used to locate the owner to within an area of twenty-five square metres. No matter where they happened to be, and regardless of what they happened to be doing (driving excepted, one hopes), the smartphone app prompted the user to report how happy they were feeling and what they were up to at the time. The results were unambiguous: those who were quizzed when outdoors were simply much happier (significantly and substantially happier, according to the study's authors) than those who responded while inside. Furthermore, various confounding factors, such as the weather, daylight, activity, companionship, time and day of the week were all controlled for with this particular experimental design.[12]

Such real-world research does, though, struggle with the question of causality. One could, for instance, imagine an alternative explanation for these results, whereby when we are feeling low, we may just simply be less inclined to get out and about than when we are in a good mood. However, when taken together with the hundreds of other, carefully controlled intervention studies that have been published, causality would seem to run mostly in the opposite direction. That is, it is exposure to nature that makes us feel better about ourselves and the world in which we live.[13] But how exactly is the nature effect explained? Is it, perhaps, simply the increased opportunity for social interaction that being in nature provides? Could it be the chance to engage in physical exercise and recreation that benefits us the most? Or is it instead, at some more fundamental level, a result of exposure to living things, both flora and fauna, as suggested by Wilson's biophilia hypothesis? According to Roger Ulrich, one of the other key figures in this area, it is recovery from psychophysiological stress that is at the core of the nature effect.

One attempt to explain why exposure to nature may actually be so good for us comes from attention restoration theory (ART).[14] According to Rachel and Stephen Kaplan, the main proponents of this particular view, when compared to urban scenes, natural environments turn out to be more effective in terms of helping restore our ability to focus our attention — what they call 'directed-attention abilities'. Basically, their idea is that nature is mildly engaging, or fascinating as some have suggested, meaning that it effectively captures our interest in a bottom-up, stimulus-driven manner.* This then allows us to replenish our top-down, or voluntary, attentional resources — this is the recovery from directed-attention fatigue (directed being another word for voluntary). These resources tend to be more actively engaged in the built environment as we try to navigate all the vehicles, pedestrians, advertisements and other distractions we encounter there and are thus in danger of being fatigued or depleted.

In support of such a view, Kaplan and his colleagues in Ann Arbor, Michigan conducted a series of highly cited experiments in which people went for a timed walk either in the local arboretum or downtown. The participants completed a mood questionnaire beforehand and once again after they had returned from their walk. They also performed various mentally taxing tasks such as repeating complex numbers backwards. A week later, the participants swapped routes, with the order carefully counterbalanced across participants. The results revealed that a nature walk selectively improved people's executive functioning — that is, their ability to make decisions and prioritize one task

---

* There is, in fact, a link to Wilson's definition of biophilia as 'the innate tendency to focus on life and lifelike processes'.

over another – while leaving their alerting and orienting functions, two other key components of the brain's core attention network, unaffected.[15] A small follow-up study conducted on twelve people showed similar effects using a different set of tasks while having the participants look at and evaluate either nature pictures or urban scenes. We will come back to this type of research in the Workplace chapter.

It is worth noting that the biophilia hypothesis and ART make somewhat different predictions about the underlying causes of the nature effect. According to the biophilia account, there is something about being amongst living things that is beneficial, while according to the Kaplans it is being in an environment that is minimally demanding of our voluntary, or directed, attentional resources that is key. There is likely to be some truth in both premises. Other researchers, meanwhile, argue that exposure to nature may also help reduce stress.[16]

So far in this chapter, we have focused on the short-term effects of exposure to nature, be it in our garden or elsewhere. But what about the longer-term consequences? Well, according to the latest findings, our brains actually change as a result of living close to certain types of nature. In order to arrive at this intriguing conclusion, German researchers scanned the brains of 341 elderly Berliners and correlated the results with the density of forest coverage within a kilometre of where they lived. Those living in a more densely forested area tended to have significantly greater amygdala integrity, meaning a higher density of grey matter, than those living in urban or green urban spaces.*[17] (The amygdala, note, is a small area lying in the middle of the brain that is involved in emotional processing.) By contrast, no

---

\* It is, though, worth noting when interpreting these findings that the higher forest density areas also happened to be on the outskirts of the city, as also, I presume, are the wealthier individuals. Causality, once again, is not as absolutely clear-cut as the authors would perhaps have us believe.

such link was found for those pensioners living in open green areas, wastelands, or else by a river.

But why, you might well ask, if nature is so very obviously beneficial for our health and well-being, don't we get out more? Even if not for a full-blown bit of forest bathing, at least sitting out in the garden more than we do, or else, better still, doing a little gardening. That, or something equally beneficial, such as a walk or jog through the local park. While part of the answer here undoubtedly relates to the steady destruction of natural habitats that has been seen in many urban areas over the years,[18] there are also limits in our affective forecasting ability that are pertinent.[19] We are simply not very good at predicting how everyday activities and/or any major life events that occur, like losing a limb or a parent, will make us feel. And predicting how good we would feel should we choose to immerse ourselves in nature turns out to be no exception. You do not need to be a psychologist, I presume, to realize that going for a walk outside in pleasant weather will probably be more beneficial for your mood than an equivalent stroll along an underground corridor. However, what we do a less good job of predicting is just how much happier we will feel. It is this ubiquitous failure of imagination that helps explain why we don't get out more, despite the benefits being so pronounced, and hence, or so one would think, obvious. In short, we simply just don't realize how much good being exposed to nature will do us.[20]

## *A room with a view: decomposing the nature effect*

Is it the sight, sound, smell or feel of nature that is most important? The evidence certainly demonstrates that the sight of nature per se (i.e. without any of the other sensory cues) can deliver a significant boost to our mental and physical well-being. In one small early study, conducted over a ten-year period in a Pennsylvania hospital, patients were found to recover more rapidly from gallbladder surgery if their room had a window looking out onto nature (deciduous foliage in the summertime) rather than a brick wall.*[21] Similarly, in another observational study, conducted at a Michigan penitentiary, the healthcare demands of those prisoners with cells facing onto farmland and forests were significantly lower than for those incarcerated in cells looking onto the courtyard. In such cases, it was presumably the *sight* of nature that was key to the positive health outcomes that were observed.[22]

The *sounds* of nature also have a beneficial effect on our wellbeing. There is certainly plenty of evidence concerning the negative health consequences of exposure to noise, defined as unpleasant auditory stimulation.[23] Replace that noise with the sounds of nature and we will no doubt all soon start to feel much better. Swedish researchers have demonstrated that people recover more rapidly from a stressful maths test in the presence of nature sounds as compared to when their ears are assaulted by road traffic noise.[24] It was certainly noticeable when Covid-19 struck just how many newspaper columnists started commenting on the little pleasures associated with the consequent lockdown, such as the birdsong outside their window no longer being drowned out by traffic noise.†[25]

What is perhaps more surprising, though, is that the greater

---

\* While this study by Ulrich, published in the leading scientific journal *Science*, has been cited more than 4,000 times in the intervening years, it is worth noting that there were only twenty-three patients in each group. This is a very small sample by today's standards. No wonder Ulrich included a 'may' in the title.
† Interestingly, birdsong has become higher in pitch in urban areas over the years as avian wildlife tries to compete with all that low-pitched urban traffic noise.

the perceived biodiversity – in this case, the more bird species that can be heard – the more beneficial/restorative nature soundscapes are.[26] Such observations certainly fit with Wilson's biophilia hypothesis. However, our primarily indoor existence means that any nature sounds that we hear these days in the office, shopping mall, airport, aromatherapy spa etc. are likely to have been piped in. Some businesses have even started to trial nature soundscapes in open-plan offices in an effort to reduce distraction and provide a sense of privacy.

We are, in other words, increasingly exposed to a simulated version of nature, which raises a crucial question to which we will return later: 'What, if anything, is lost when we attempt to reproduce nature in the mediated sensorium?' This issue becomes absolutely key when architects, designers and others try to capture the benefits of nature by presenting the pre-recorded sights, sounds and even synthetic smells of nature. However, there is, I think, a very real danger that the sounds of the forest in the airport, or the noises of the jungle in the toy store (both examples that we will come across in later chapters), may just be treated as incongruous by whoever hears them. Under such conditions, what we hear bears little or no obvious relation to what we are seeing, nor probably to what we are smelling either. Such incongruent combinations of sensory inputs can be hard for us to process, and may well be negatively valenced as a result, meaning that we do not like them.

When considering the beneficial effects of *smelling* nature, one is soon drawn into the realms of aromatherapy. Just think of all those essential oils such as lavender, citrus, pine and peppermint that can benefit our health and well-being. While you may be tempted to associate a number of these scents with food, they are more fundamentally the smells of nature. At the same time, it is important not to forget that nature can also smell awful; just take the manure-stink of the farmyard. It is more than a little ironic that while our brains tend to adapt to, and hence lose awareness of, those smells that we classify as either pleasant or neutral (like the BO of our home), we never seem to adapt to those smells that we consider unpleasant. This is especially unfortunate for anyone living close to a battery farm or rubbish dump. According to the scientists, our brains tag unpleasant smells as potentially dangerous. Hence, we retain our awareness of them, so that we can continue to monitor their source, in a way that is simply not the case for pleasant or neutral smells that can safely be ignored once our brains have classified them as harmless.[27] Not much comfort for those who are so afflicted, I am sure, but at least now you know.

We should not forget to at least mention the *taste* of nature. Think only of the traditional physic garden, such as one finds in London's Chelsea,[28] or David Mas Masumoto's elegiac ode to the peaches he farms in California in his 2003 book *Four seasons in five senses: Things worth savoring*. As the anthropologists have pointed out, 99 per cent of human history was lived exclusively as hunter-gatherers. Relative to that, agriculture has lasted for no more than a moment, and the urban landscape for little more than the blink of an eye.[29] Ultimately, therefore, the nature effect presumably originates in an innate drive that encouraged our ancestors to inhabit those environments likely to provide both sustenance (hence the taste of nature) and safety.

Which just leaves the *feel* of nature. Surprising though it may sound, touching plant foliage (not to mention tree-hugging) can exert a positive effect on the mood and well-being of both children and adults.[30] A regular dose of gardening can be a great way to make contact with nature, even if you're stuck in an inner city. Gardening anywhere is, of course, also a mindful activity, one that can make us pay attention to the sensory properties of the nature that surrounds us, and how they change with the seasons. Indeed, gardeners often talk of the 'fascination' that their contact with the natural world evokes. The suggestion is that this stimulus-driven capture of attention as we watch plants grow may provide the breathing room needed for our directed attention to recover.[31] While thinking about the touch of nature, we should perhaps also stop to consider the feel of the breeze against our skin, the warmth of the sun on our back and the rain on our face. Not only can we touch nature, in other words, but nature also touches us right back.

## Ranking nature's benefits

Having taken a look at what each sense contributes individually, as far as the nature effect is concerned, I now want to turn to address the question of whether there is some kind of sensory hierarchy in operation. In the opening chapter, we came across the notion that humans are visually dominant. So does that mean that the beneficial effects of viewing nature necessarily trump those of listening to it? Is it better to smell nature than to feel it? As yet, we simply do not have satisfactory answers to such questions. Anyway, this is actually a rather tricky question to address properly; comparing one sense with another is a bit like trying to compare apples with oranges. However, the more important, or at least tractable, question for me is whether the benefits of the nature effect can be enhanced if more of our senses are engaged in the right way. Is having the sight and sound of nature more effective than having the best of the individual senses? Similarly, one might want to know whether the pattern of multisensory stimulation, be it what we see, hear, feel and/or smell, has to be congruent, or if that perhaps does not matter?[32]

To put things more concretely, what do you think would be the outcome for someone listening to some drum and bass or death metal over their headphones while simultaneously walking through a forest or park? Does the throbbing soundtrack neutralize, or in any way interfere with, or diminish, the visually induced nature effect?* And what should we say about all those people with headphones on, or else glued to their smartphone screen all day long? Do they still benefit from nature? These are precisely the sorts of questions that came to mind whenever I cycled through the English countryside on the way to visit my mother in a nursing home a few miles out of town. In this case, it was the traffic noise and exhaust fumes from the main road that I worried about. Did all that noise and pollution negate any benefits that might have accrued from the beautiful countryside I was cycling through? While no one has addressed quite this question, the research clearly suggests that vehicle noise impairs people's appreciation of, and memory for, visual scenes.

A number of studies have demonstrated that those viewing pictures of national parks in the US exhibit an impaired memory for, and a much reduced appreciation of, those landscapes if they also hear road traffic noise, or else the sounds of helicopters – this being the cacophony that apparently accosts the ears of many of those who are lucky enough to visit the Grand Canyon.[33] In one study, for instance, recreational motorized noise reduced people's aesthetic appreciation of the landscape by as much as 30–40 per cent from when only the sounds of nature were audible. Most of us find the sound of motorcycles especially disturbing. A correlation has been documented between the diversity of recorded birdsong played and people's appreciation of static urban landscapes projected onto a screen.[34]

---

* Or, worse still, country and western, given evidence linking suicide rates in the US with the number of hours of this kind of music playing on the radio, at least according to Stack and Gundlach (1992).

The key point to bear in mind here is that our senses interact *all* the time. As we have seen already, what we are taking in via any one of them can very definitely influence, enhance and/or mask whatever may be going on in the others. There is no reason to believe that our perception of/response to nature is any different in this regard. In the chapters that follow, we will come across many more examples where our response to specific environments can be understood only by assessing what is going on in several of our senses simultaneously. At the same time, however, what we believe about the source of that nature stimulation also matters: is it really nature that we are seeing, hearing, smelling or feeling, or just a digital/synthetic reproduction?

## *Natural timing*

Nature helps train our bodily rhythms. Spend too much time indoors, with an essentially constant pattern of multisensory stimulation – ambient light that is as bright, or dim, first thing in the morning as it is late in the day – and people soon start to suffer. What we miss are, in part, the synchronizing sensory cues, such as the diurnal changes in ambient light that help keep our internal rhythm synchronized with that of nature. Hence the recommendation to get a dose of natural sunlight to overcome jet lag on reaching some far-flung destination.* The hue, or colour, of natural light also changes predictably over the

---

* At the start of my academic career, I worked on the Crew Work Station of the European Space Agency. I often read about research concerning what happens to astronauts confined for months on end to a cave or in the far north of Scandinavia in the middle of winter, with no reference to the natural diurnal cycle. The evidence from such research shows that without any external reference, our internal clocks run on something like a 22.5-hour cycle, not a 24-hour one, as you might have expected.

course of the day. Photographers know only too well that people look better in the warm golden yellow light of the late afternoon than in the colder bluer hues of early morning.[35] Indeed, mimicking the blue of dawn by means of coloured interior lighting can provide a surprisingly effective means of enhancing people's alertness, not to mention reducing suicide rates.[36] Given that we have evolved in environments that changed systematically over the course of every day, perhaps the mediated sensorium in which we spend so much of our time should be designed to imitate this. This is where getting out into the garden can really help.

But what of the natural rhythms of the other senses? I, for one, often wonder about the dawn chorus, especially when the birds start chirping and tweeting at 4.30 a.m. as it gets light in the summertime. It certainly feels like that racket coming from the large tree outside my bedroom window is having much the same effect on my alertness and arousal as a dose of blue light, much though I sometimes wish that this was not the case. Given that the make-up of the birdsong in the morning is distinct from what can be heard at dusk, that will presumably also make a difference in terms of how alerting such nature sounds are (i.e. it is not just the intensity, loudness or brightness that matters, but also the type of stimulation). What might the sound of the rooster's cock-a-doodle-doo do to us? That is an experiment the results of which I would like to see.

Nature also smells more or less fragrant depending on the time of day: more so during the day and early evening, somewhat less so at night and first thing in the morning. This is why plants that release their intoxicating perfume only once the sun has gone down, such as the caballero de la noche (otherwise known as *Cestrum nocturnum* or night-blooming jasmine), stand out amongst the generally olfactorily neutral backdrop of the relatively odourless, at least to us humans, cool night air. Once upon a time, British landowners would build walled gardens outside their manor houses, in part to help capture and retain the gorgeous scent of all the plants and flowers that they, or more likely their gardeners, were cultivating within. Nowadays, however, there is little point, given that so many of our plant varieties have been selectively bred to look good (i.e. colourful). Unfortunately, this means that much of the scent is lost, as there is normally a trade-off between looking good and smelling good. Most plants invest their precious resources in advertising themselves by one or other route, not both.* Of course, many people would say that we face exactly the same problem with so much of our beautiful, but tasteless, supermarket fruit and veg.

In her 1995 book *Worlds of sense: Exploring the senses in history and across cultures*, Constance Classen charts the evolution of the sensory landscape over time through an analysis of our attitudes toward roses. By scouring the works of Western writers, poets and gardeners, across the centuries, she finds that while early writers such as Pliny focused their descriptions of roses primarily on their scent, modern writers tend to emphasize their colour and other visual attributes. Just take the following from Alicia Amherst, author of *A history of gardening in England*, published back in 1896, who put it thus: 'a rosery of to-day would astonish the possessors of gardens in the Middle Ages, and the varied forms and colours would bewilder them, yet in some of our finest-looking roses they would miss, what to them was the essential characteristic of a rose, its sweet scent!'

---

* Wild orchids (i.e. not those that have been bred for the florist's shop) seemingly being one of the few exceptions in this regard. Coincidentally, this flower was chosen as the cover picture for Kellert and Wilson's (1993) book *The biophilia hypothesis*.

My home in the Colombian cloud forest.

## Santandercito, my garden retreat

Nowadays I, like many people, seem to spend most of my time in front of a computer screen or else stuck in traffic commuting from one conference or meeting to another. However, I am married to a Colombian and we are lucky enough to have inherited a small *finca* in the mountains just outside of Bogota. In this cloud forest environment, extreme gardening is the order of the day. And while part of my time is spent tackling the weeds, which grow with surprising vigour and tenacity – that is the extreme part of the gardening – we also have plenty of time to cultivate our own fruits and herbs. In the former category, many kinds of confused citrus (more mixtures of lemons, oranges, limes and mandarins, in fact, than anyone would have thought possible), some phenomenally fiery chilli plants that I must admit to being inordinately proud of, as well as a number of other exotic fruit species. For weeks on end, I don't leave the front gate. Very Peter Mayle, I know. However, the beneficial effects of looking at and listening to as well as smelling, feeling and tasting nature are so powerful that you can really notice them working their magic on you. The way I write, even the way I think, changes as the days immersed in this tropical garden paradise start to add up; remember the dose-dependency we came across earlier?

At the same time, however, the lush tropical cloud forest of Colombia undoubtedly constitutes a very different kind of environment from that of rural England. So the question arises as to whether one type of nature is, in some sense, better than another. Is the seaside better than the forest, or ornamental gardens? Do botanical gardens or arboretums serve a more restorative function than the wild copse, savannah or jungle? And is the Colombian cloud forest really any more restorative than, say, Port Meadow, the flood plain that lies close to our home in Oxford? These are, of course, precisely the sorts of questions that need answering if one wants to sensehack the nature effect in the built environment. That being said, one might also need to be sensitive to cultural differences if Japanese writer Jun'ichirō Tanizaki is correct when he writes in his essay on aesthetics *In praise of shadows* that the Japanese like their gardens well stocked with dense plantings, whereas in the West the preference is more often for a flat expanse of grass.

Fortunately, researchers have already started to compare the effects of well-organized versus more naturalistic spaces. Broadly speaking, their findings support the view that the more informal the garden layout you have, the better.* The research also shows that people prefer the forest or tundra to desert or grassland scenes. Botanical gardens score somewhere in the middle.[37] Of course, according to the biophilia hypothesis, one might well

---

* Nineteenth-century English and North American 'wild gardens', twelfth-century Japanese stroll gardens and late-eighteenth-century English landscape gardens would all fit into the 'informal' category, such as the wilderness created at the back of St John's College, Cambridge, where I was once lucky enough to be a junior research fellow.

have predicted that botanical gardens and arboretums ought to have been *more* restorative, delivering a more pronounced benefit, given that, if well stocked, they should presumably offer the visitor far more biodiversity than one is likely to come across in a native woodland. At the same time, however, one also needs to factor in crowd density, as that can offset the beneficial effects of being in nature. What is more, I suppose there may be a sense in which the sound of nature (the fauna, that is) is largely absent from most botanical greenhouses. Finally here, we have the question of congruency – does the sound of the fauna need to match the sight of the flora? The research suggests that it should, though too often, of course, it doesn't.

However, while extolling the virtues of nature, it is, of course, important not to forget that the fauna and flora can also be, respectively, extremely dangerous or downright toxic.[38] (Here I am thinking of the many tarantulas, the very occasional deadly snake and the plants whose leaves bring my skin out in unbearably itchy rashes for weeks on end after casually brushing against them in my garden in Colombia.) No surprise, then, that our 'innate' affinity for nature (biophilia) is matched by biophobia, an equally strong fear of things that look like, or move like, or, indeed, are spiders and snakes, etc. Biophobia appears to reflect some kind of evolved preparedness to develop a fear of, or acquire an aversion to, these ancestral threats.[39] While so many of our sensory responses are mediated, learned and/or context dependent, it is tempting to believe, in line with the title of Diane Ackerman's best-selling book *A natural history of the senses*, first published in 1990, that the fundamental response to life, to certain kinds of nature at least, be it of the biophilic or biophobic variety, is based on tendencies that are innate.[40]

While an evolutionary account of the nature effect (stress recovery theory) is often put forward, it is worth noting that an alternative account in terms of perceptual fluency might well be as plausible. The idea here is that we find it especially easy to process natural scenes because their fractal arrangement matches the statistics that our visual systems evolved to process.[41] According to Joye and van den Berg, 'The fractality of natural environments and elements is – among others – clear from the fact that such shapes/scenes consist of increasingly smaller copies of themselves over a large number of scales of magnitude.'[42] For example, just think about a tree: all of the branches – from the largest to the smallest – are scaled-down versions of the entire tree. What this means, in practice, is that one part of a natural scene already provides us with an idea of what is likely to be in other parts of the scene. In other words, natural environments are often characterized by a deep degree of perceptual predictability/redundancy. This results in increased processing fluency. By contrast, urban scenes often tend to consist of stimuli that appear very different, one from the next, with the different stimuli all competing for our attention. This makes it much harder for us to figure out the gist of a scene. That hypothesis would certainly appear to be consistent with the rapid speed with which people process/understand natural scenes. Given that processing fluency is usually associated with positive affect, as Reber and colleagues have documented in a number of publications over the last twenty years, this could provide an explanation for why we are drawn to such unthreatening natural scenes.[43] The notion of processing fluency is one that we will come back to time and again in the chapters that follow.

## *Sensehacking nature for well-being*

There are undoubtedly many more important questions to be addressed here regarding the causes and consequences of the nature effect. However, the key point to bear in mind for now is that any consideration of sensehacking has to begin with our response to the natural environment, and to the patterns of multisensory stimulation that we have evolved to deal with over the millennia. The more that the indoor environments in which we spend so much of our time working, exercising, playing, shopping and relaxing deviate from the natural ideal, the less happy and productive we are likely to be. No matter whether or not you buy into the biophilia hypothesis, a mounting body of scientific evidence now demonstrates that exposure to nature, even in short bursts (or should that be doses?), can have a remarkably restorative effect on us (and this, based on the evidence, is not an overstatement), on our mood, our health and our well-being too.[44]

Experiencing nature undoubtedly can help improve mood, while at the same time reducing cognitive fatigue, rumination and stress.[45] Some commentators have gone even further, extolling the more spiritual benefits of the mindful multisensory communion with nature.[46] The key point to remember is that we experience nature through *all* of our senses, and so ensuring as balanced a diet of natural multisensory stimulation as possible is probably one of the best things that any of us can do to promote the health and well-being of both ourselves and those we care most about. And for those of us who are lucky enough to have one, getting out into the garden can be a great way in which to tap into these benefits.

Bear in mind here, though, that it is not enough simply to *know* that you are close to nature, you really have to *experience* it with all of your senses to get the full benefit. Anyone who disagrees should just consider how they would feel if their holiday hotel accommodation was not one of the brochure rooms with a sea or countryside view, but rather one facing the urban sprawl inland.*

---

\* A scenario central to E. M. Forster's 1908 novel *A room with a view*.

No matter whether it is big or small, or if we live in a climate tropical or temperate, it is important to recognize that any outdoor space provides one of the most important opportunities to lower our stress levels and improve our well-being. No wonder, then, that many national papers in the UK were endorsing the benefits of growing-your-own during the coronavirus lockdown. But wherever you connect with nature, be it in your own back garden, or the local park or forest, just make sure to enjoy it with as many of your senses as you can. The benefits for your social, cognitive and emotional well-being will be well worth the effort, whether you realize it or not.

# 4. Bedroom

Irrespective of whether you think it a waste of time or the highlight of your day, sleep takes up more of our time than pretty much any other activity. On average, we spend something like a third of our lives sleeping, or at least trying to. While many of you have undoubtedly heard of those legendary characters like Margaret Thatcher and Ronald Reagan who could apparently operate effectively on as little as four or five hours' sleep a night, it may well be more than mere coincidence that both world leaders subsequently went on to develop dementia in their later years. Indeed, the evidence now suggests that we all need a regular seven to eight hours of quality sleep a night (no matter whether we want it or not), if we are to avoid increasing our risk of a whole host of negative health outcomes. These include everything from dementia to obesity, and from cancer to heart disease. Insufficient sleep has been linked with seven of the fifteen leading causes of death, including cardiovascular disease, cerebrovascular disease and diabetes.[1]

Sleep research is big business these days. No surprise really, given the suggestion that between one- and two-thirds of those living in North America and Britain fail to get enough sleep on a regular basis, with similar problems reported in many other industrialized nations.[2] In fact, insomnia has been defined as the second most common psychological disorder after chronic pain, with a prevalence currently running at something like 33 per cent, with 9 per cent of people reporting that their sleep difficulty occurs on a nightly basis.[3]

According to the popular press, we are more sleep deprived

than ever before. Many eminent sleep scientists have also vociferously pushed the same line. For instance, Matthew Walker, the Liverpool-born researcher currently based in California, and author of the international bestseller *Why we sleep*, argues that 'the shortage of sleep has reached epidemic proportions', and talks of a 'catastrophic sleep epidemic'. According to Walker, less than 8 per cent of us were trying to survive on six hours or less back in 1942, as compared to one in two of us in 2017.

## The less you sleep, the shorter you live

The negative health implications for anyone who sleeps for less than six hours a night are frightening. According to a 2016 report from the RAND Corporation, they are 13 per cent more likely to die at any time than those who get seven to eight hours of sleep. If you are at least forty-five years old and get less than six hours of sleep a night you are 200 per cent more likely to have a heart attack or stroke during your lifetime.[4] Writing in top science journal *Nature*, Till Roenneberg has suggested that we are getting between one and two hours' less sleep a night than our forebears did fifty or a hundred years ago. No wonder that the Centers for Disease Control and Prevention (CDC) in the United States have gone so far as to declare insufficient sleep a 'public health problem'.[5] Put the evidence together and it is easy to understand why the financial cost to society of insomnia-related problems is so dramatic, estimated at several per cent of gross domestic product, according to the report from the RAND Corporation (e.g., 2.92 per cent of GDP in Japan, equating to $138 billion; 2.28 per cent of GDP in the US, equating to $411 billion; and a not inconsiderable 1.86 per cent of GDP in the UK, equating to $50 billion).[6]

A one-hour increase in the weekly mean amount of sleep is

associated with a person's earnings increasing by 1 per cent in the short term and almost 5 per cent in the long term. Meanwhile, the results of Vitality's 2019 'Britain's Healthiest Workplace' study revealed a correlation between income and sleep quality, with 57 per cent of those on less than £10,000 a year reporting sleep problems as compared to just 23 per cent of those who earned more than £150,000. That said, causality is uncertain.[7]

The growing realization of just how important a good night's sleep is to our social, emotional and physical well-being has led to the emergence of concepts such as 'sleep hygiene' and 'sleep engineering'. Regardless of the terminology, the basic idea here is to try to establish the right evidence-based practices to help those who may be struggling to get the necessary amount of the right kind of sleep. Bear in mind that we sleep in cycles of roughly ninety minutes, including non-rapid eye movement sleep, followed by rapid eye movement (REM) sleep (where everything but our eyes are paralysed) and then slow-wave sleep. It is the last one you really need to make sure you get enough of in order to consolidate memories.

Simple fixes include everything from more exercise to sex.* According to the experts, we should not eat for at least a couple of hours before we sleep, as much for the sake of our waistline as for enhanced 'sleep hygiene'. Drinking alcohol in the evening lowers the quality of our sleep too, as does ingesting caffeine or other stimulants.[8] Most people already know about many of these suggestions, though, for whatever reason, choose to ignore them. Hence I want to take a look at a number of sense hacks that may help us to sleep better. Indeed, there are various sensory tricks that we can all use to help us reduce the amount of

---

* Or, as the scientists at Harvard Medical School dryly put it when offering their top tips for a good night's sleep: 'It may help to limit your bedroom activities to sleep and sex only.'

time before we fall asleep, improve our sleep efficiency and wake with a little more of a spring in our step.

Perhaps counterintuitively, one of the behavioural techniques for helping people to sleep better is simply to restrict their sleep for a while. Researchers in Oxford have developed a sleep-enhancing app called Sleepio, which is proving immensely helpful to many people to retrain their sleeping patterns and behaviour. The app helps users to log their sleep patterns while at the same time providing a structured sleep schedule, limiting the amount of sleep, at least to begin with. The app has proved so successful that it is now being rolled out across sections of the National Health Service. Indeed, cognitive behavioural therapy has been shown to work well in the long term in treating the symptoms of insomnia. Hypnotic drugs, known as benzodiazepine-receptor agonists, can also be used in the short term. That said, the side effects of using drugs such as temazepam, which I was once prescribed, include addiction, memory problems, male breast growth and birth defects. I guess I was lucky only to end up with the moobs! What is more, sleeping medications tend to encourage the wrong sort of sleep, that is, they do nothing to enhance slow-wave sleep. And if that wasn't enough to put you off, they may cause cancer too.[9]

The sleep-health industry was worth £30 billion a year in 2018, with that figure estimated to more than double by 2020. There has been an explosive growth in companies promising to improve the quality of your sleep by offering a host of more-or-less plausible neuroscience-inspired sense hacks.

## *Nodding off*

Have you ever tried counting sheep when you were struggling to get to sleep? If so, I am sorry to be the one to have to tell you that you were, quite literally, wasting your time. At least, you were according to research from one of my former colleagues in Oxford, Professor Allison Harvey. She has long since disproven this old wives' tale. When Harvey had a group of insomniacs count sheep, or otherwise suppress negative thoughts while they were nodding off, sleep onset was delayed by an average of ten minutes. So what should you be thinking about instead? Well, conjuring up a tranquil and relaxing scene such as a waterfall or being on holiday helped people in one study to get to sleep around twenty minutes faster than those who were given no specific instructions. The explanation in this case was that maintaining the pleasant mental imagery was sufficiently cognitively demanding to prevent the participants, who all suffered from insomnia, from ruminating over negative or worrying thoughts.[10]

## Dazzled by the light

Do you sleep with, or next to, your phone? Bad idea. According to the results of a 2015 survey of 1,000 North Americans, 71 per cent of us sleep with, or next to, our smartphones: 3 per cent sleep with it in our hand, 13 per cent keep it on or in the bed and the other 55 per cent have it within easy reach.[11] When trying to tackle the problems that so many of us face with nodding off, it is vital to realize that the sensory aspects of the environment may well be partly responsible. Indeed, these days, blame is often placed on our growing exposure to artificial light at night, much of it coming from increased screen time in the evening hours. According to a 2015 study,[12] those who read an electronic book on a light-emitting e-reader in the hours before bedtime take longer to fall asleep, feel less sleepy in the evening, secrete less melatonin,* exhibit a later timing of their circadian clock and also show less next-morning alertness than those who read a print book.† Such findings are especially concerning given that more than 90 per cent of us use some type of electronic device at least a few nights per week within an hour of going to bed. Indeed, given the growing awareness of the dangers of excessive screen time, some commentators have even started to question whether the device manufacturers themselves may have a responsibility to start doing something to protect us all against the blue light they give off – blue light being especially bad as it can trick our brains into thinking it is time to wake up.‡[13]

Warning: sleeping with the lights or TV on is associated with an increased risk of weight gain and obesity. This according to a cohort study published by the National Institute of Environmental Health Sciences in North Carolina. These researchers followed more than 43,000 women aged between thirty-five and seventy-four years over a five-year period. Compared to those sleeping without artificial light, those exposed to light at night were, on average, 5kg (11lb) or more heavier.[14] It appears, then, that artificial light disrupts or delays the body's natural clock and upsets the normal hormone balance. Even though the visual stimulation didn't have any noticeable effect on self-reported sleep quality, and despite the fact that causality cannot be determined, these results

---

* Melatonin is an important hormone secreted by the pineal gland that helps control the sleep–wake cycle. During the day, levels are low, but start to increase once the sun goes down.
† That said, it is worth noting that the participants had to read for four hours straight with the light-emitting device turned up to its maximum brightness.
‡ The blue light of dawn having a slightly different wavelength than that seen at other times of day.

nevertheless do hint at the potential benefits of removing as much light as possible at night, both before and after we go to sleep. Sensehacking, then, may be as much about removing unwanted sources of environmental stimulation as it is about adding new ones.

In terms of concrete sense hacks to help reduce your exposure to artificial light in the evening, you could try some of the following suggestions. Most importantly, avoid looking at bright screens for two or three hours before you go to sleep, exercising what is known as a cyber curfew. If however, like me, you are obliged to use various electronic devices in the evening, either wear purpose-made glasses or else download an app that will filter out the blue light. Some mobile devices have a night-time setting that automatically shifts the display screen to warmer colours once the sun sets. If you need light in the bedroom, dim red night lights can help as they typically do not affect our melatonin levels as much as do other light colours.[15]

The notion that we should all be reducing our exposure to light at night would seemingly run counter to the idea behind the Dodow, a commercial device that emits an expanding and contracting ring of blue light when pointed at your bedroom ceiling. In this case, those with problems sleeping are encouraged to try to synchronize their breathing, inhaling when the circle expands, and breathing out when it contracts, in an almost meditative manner. However, while one can find plenty of journalists who have tried the Dodow, there isn't yet any good scientific evidence, at least as far as I can see, to show that this gadget actually helps you to sleep better. What is more, the choice of short wavelength blue light would seem like an especially bad idea, though according to those behind the product, the light is supposedly too dim to keep you awake.

## *Sweet dreams?*

I have spent a surprising amount of my research career contemplating the multisensory design of toothpaste. One of the things that has always struck me as strange is how we mostly use the same formulation morning, noon and night. It would seem to me that both our physiological and psychological needs are quite different at either end of the day. After all, just think about how we eat and drink very different things for breakfast and dinner. Caffeinated drinks in the morning and chamomile tea, or some other herbal infusion, before bed. This despite the fact that there is surprisingly little evidence supporting the sleep-enhancing properties of chamomile tea.[16] Some of you may be all too aware of how skin creams come in different variants for day and night as well. So why not different toothpastes too? I have little doubt that the invigorating and alerting aroma of mint is just what we need to perk us up and help avoid sleep inertia first thing in the morning. At the end of the day, though, when we are trying to wind down and ready ourselves for sleep, surely this is the last thing we want?*

## *Sleeping soundly*

How often have you found unwanted noises keeping you awake at night? A major problem for many of us is environmental noise exposure, especially for those living close to airports or other forms of transport, leading as it does to increases in obesity and mortality. According to the World Health Organization,

---

* In some parts of the world, toothpaste comes in a variety of flavours, including orange and liquorice.

disturbing levels of ambient noise result in the loss of more than a million healthy life years annually in western Europe alone. Most of these can be attributed to noise-induced sleep disturbance and annoyance. It is not just when you are trying to get to sleep that noise can be problematic; it can even be a problem when you are already asleep. In fact, there is even a suggestion that night-time noise may have a more detrimental effect on our cardiovascular well-being than equivalent noise levels experienced during daylight hours.[17] One of my favourite tricks here is to use nature sounds, such as waves gently breaking on the beach, to help mask unexpected background noise at night. Or, when that isn't available, the white noise from a mistuned radio does the trick admirably (some companies already sell white-noise generators for just this purpose).

On occasion after using this radio sense hack, I have woken the next morning hearing nothing but silence and thinking I must have turned off the radio during the night. Then, after a few seconds, the sound of the white noise suddenly leaps back into aural focus.* Talking of which, for those living next door to someone with a loud radio, try tuning your radio to the same station and matching the loudness. While this will not change the decibel level, you, or rather your brain, should find it just that little bit easier to ignore a quiet radio close by than a loud one further away. While this might sound like a crazy idea, I believe that this curious sense hack is at least worth a try. That said, given what we saw earlier about the increased weight gain in those who fall asleep with the TV on, you might be well advised to buy a radio with a 'sleep' function that will switch it off once you have fallen asleep.

* The explanation for this peculiar phenomenon is presumably that our brain tends to tune out constant sources of stimulation. It normally takes a few seconds for our attention to start working properly when we come round first thing in the morning.

Other bizarre solutions designed to help you sleep that have hit the headlines recently include a recording of John McEnroe reading *The rules of tennis: a love story* or *But seriously: the rules of tennis*, available via the Calm mobile phone app. I don't know about you but I think that I would rather listen to Pzizz – an app that plays a blend of natural sounds, soothing music and guided meditation. Another growing trend is for people to access some autonomous sensory meridian response (ASMR) content online from the likes of actresses such as Eva Longoria and Margot Robbie. The shiver that some people experience running down the back of their neck when someone whispers and meticulously crinkles paper, supposedly helping them to relax gently into sleep. There are several eight-hour loops of nature sounds on YouTube too.

## *Sleep on it*

I am sure that you have come across the expression before, but does sleeping on a problem really help? In some intriguing research published in *Nature*, sleeping increased the likelihood of people solving an insight problem more than an equivalent period of time during the day, or a nocturnal period of wakefulness. This is what is referred to as the 'ah-ha' moment. In fact, creative solutions were up threefold following a good night's sleep. What is more, we can also learn new contingencies between sounds and scents while we sleep. Furthermore, presenting sounds and scents associated with stimuli that were learned during the day while in slow-wave sleep helps people to consolidate their memories. However, before getting too excited by all these results, it is worth bearing in mind that the nature of the learning involved in these studies was pretty simple. So, as yet, there is no evidence that you can, say, learn a foreign language while you sleep, though some limited ability to improve your vocabulary is certainly not beyond the realms of possibility.[18]

## Sleep tight

Taking a hot shower or bath (foot or body) an hour or two before bedtime, even for as little as ten minutes, helps those struggling to sleep to nod off sooner.[19] Passive body heating (PBH), as the authors of one systematic review and meta-analysis call it, also has a positive impact on sleep quality, increasing slow-wave sleep. The ideal water temperature for your evening bath is 40–42.5°C (104–108.5°F). Get the timing right, and you can expect to nod off up to 8.6 minutes, or 36 per cent, sooner than you otherwise would. Taking a warm bath or shower helps to redirect circulation to the hands and feet, thus triggering a drop in core body temperature. The body's internal clock then takes this as the cue that it's time for sleep. Note that our body temperature naturally declines in the run-up to bedtime and continues to do so as we sleep, reaching its lowest point at around 4 a.m. Hence anything that you can do to help lower your core body temperature, preferably by about 1°C, is likely a good idea, since this is the key parameter driving sleep onset. This is where lying down helps as this change in posture encourages the dissipation of heat from our core to our extremities.

'Warm feet promote the rapid onset of sleep', or so proclaims the title of an article in *Nature* claiming that warming your feet is a surefire way to decrease how long it takes you to fall asleep. The reason this particular sense hack works is that the hands and feet, along with the head, are the skin sites where thermoregulation by means of increased arterial blood flow is most effective. Practically speaking, then, if you like to go to bed with a hot-water bottle at night, you will probably get off to sleep quicker if you place it under your feet rather than hugging it close to your chest. At the other end of the body, you could try the Moona thermo-regulated pillow with associated app. And for something a little more high tech, how about the Somnox Sleep Robot, a heavy, kidney-bean-shaped cushion that 'breathes' slowly and which you are supposed to cuddle?[20]

The ambient air temperature is also an important factor in terms of sleep quality. If, for example, it increases to 30°C (86°F) then sleep quality declines. You can expect to get thirty minutes' less shuteye if the temperature in the room where you sleep increases from 18°C to 25°C (64°F to 77°F). So putting all the research together, what you are ideally looking for if you want to stand the best chance of slumbering soundly is a quiet, dark, cool environment. You should aim to keep the temperature comfortably cool (16–24°C; 61–75°F), and the room well ventilated. Minimize any auditory distraction with earplugs or 'white noise'. Use heavy curtains, blackout blinds, or even just an eye mask to help block out as much of the light as possible, since this is one of the most powerful sensory cues telling your brain it's time to wake up.[21] The one other tip that has been documented to improve sleep recently is rocking, as in the children's lullaby 'Rock-a-bye baby'.[22]

## *Why not pop a pot plant on your bedside table?*

A pot plant on the bedside table can help you to sleep. In particular, certain varieties of houseplant can help you to beat colds and tight chests, and even fight insomnia.[23] English ivy, for instance, helps to tackle airborne mould, eliminating most of it in a matter of just a few hours. Meanwhile, according to NASA, aloe vera is one of the best plants for air purification, because it releases oxygen all night long while at the same time helping to absorb and break down airborne pollutants such as benzene, a component of many detergents and plastics, and formaldehyde, which is found in many varnishes and floor finishes. And, should you happen to come across a Madagascan areca palm, then you might like to know that it came out top in terms of mopping up airborne pollutants. It also releases moisture into the air, thus potentially helping those with a cold or sinus problems to breathe a little easier.[24] The importance of tackling the problem of poor indoor air quality is emphasized by the conservative suggestion that a minimum of 99,000 deaths a year in Europe alone are attributable to its negative effects.[25]

## Are you an owl or a lark?

Are you one of those people who likes to stay up late and not wake up before 10 or 11 a.m., or do you prefer to go to bed early and get up with the lark? There is an increasingly well-recognized individual difference in the pattern of sleep that we prefer. Around 30 per cent of the population are night owls, 40 per cent are larks, and the rest fall somewhere in between. Larks tend to have slightly faster internal clocks than owls. Researchers working with the UK Biobank and the gene-testing site 23andMe assessed almost 680,000 people, 86,000 of whom had their sleep timing measured with an activity monitor. Using genome-wide data, 351 genetic loci associated with being a lark were identified. The 5 per cent of larks carrying the most 'morningness' alleles were shown to wake up an average of twenty-five minutes earlier than the 5 per cent carrying the fewest.[26] The bad news for night owls is that staying up late is linked with a range of health issues, mood disturbance, poor performance and increased risk of mortality.

Some researchers have been investigating whether they can hack the senses of night owls in order to align their sleep-wake cycles using a methodologically robust randomized control trial experimental design. In fact, simply by using targeted exposure to light, by setting earlier waking and sleep times, carefully fixing mealtimes, caffeine intake and exercise regime, it proved possible to reprogramme twenty-two night owls, advancing their sleep-wake cycles by an average of two hours, without any loss of sleep.[27] It will be an interesting question for future research to determine whether such sensehacking also helps to reduce mortality rates in the long term, assuming, that is, that the night owls can be convinced to stick with the programme. It is easy to understand why big pharma have been so keen to get their hands on 23andMe's data in order to try to develop new sleep medications based specifically on the body's circadian clock.[28]

## *The first night effect*

If you are anything like me, then you always struggle to sleep on the first night in a new place. This is a real pain as I have to spend a few days most weeks on the road 'spreading the word' in cities far and wide (at least, I did before Covid-19 struck). The bed may be super-comfortable and the hotel as fancy as you like, but still there is just something about the unfamiliar setting that relentlessly keeps me awake. This is actually a well-known phenomenon called the first night effect (FNE). While researchers first identified it more than half a century ago, it is only recently that they have finally started to figure out what is actually going on. According to the sleep scientists, one side of our brain stands guard like a night watchman whenever we find ourselves trying to sleep in unfamiliar surroundings. In this regard, we are much like marine mammals and birds – think dolphins and ducks. They sometimes sleep with just one side of their brain too. Mallard ducks, for example, are more likely to keep one eye open and sleep with the other half of their brain, as the risk of predation increases.[29]

No matter how jet-lagged we might be, we are destined to suffer from the FNE whenever and wherever we check in to a new hotel, or else perhaps stay with friends for the first time. This evolutionarily ingrained response obviously serves little purpose nowadays. It is, after all, highly unlikely that there will be a predator lurking in the closet, no matter whether one is staying in a Travelodge or the Shangri-La. The question remains, however, as to which sensory cues our brains use in order to decide that the surroundings are unfamiliar. Is it the scent, or perhaps the unusual noises, doors banging, pipes clanging, that sort of thing? One sense hack here involves trying to replicate some of the sensory cues that we find in our own homes. So, for example, if you use an air freshener or bed-linen scent at home, take some with you when you travel. You might just find that it helps you to hack your senses and convince your 'night watchman' to lower its vigilant guard a little. Another suggestion for those returning often to a location is always to stay in the same hotel, and preferably in the same room (especially if the walls happen to be painted a soft shade of blue, which according to a survey of 2,000 Travelodge Hotels in the UK was the room colour that best helped their guests to get a good night's sleep).[30]

A sense hack that many people already use is to wear earplugs to block out background noise. My suggestion here would be to fit the earplug especially snugly into your right ear. Why? Because it is always our brain's left hemisphere that stands watch, at least initially, while our right hemisphere slumbers. Hence, in order to sensehack your sleep, you need to minimize any sensory disruption to the hemisphere that is keeping guard so that it does not wake you up quite so often. So why, you might well be wondering, did I suggest making sure the earplug fits snugly in your *right* ear when it is your brain's *left* hemisphere that is keeping guard? The reason is that most of our senses project contralaterally. Whatever you feel on one side of your body, or the sounds that enter one ear, are actually processed, at least initially, in the opposite hemisphere of your brain. The only sense that does not project contralaterally is the evolutionarily much older sense of smell. Here, the left nostril projects straight back to the ipsilateral left side of the brain. So, now you know!

Should the earplugs and air freshener not help you to get more shuteye when you next find yourself staying in a new venue, rest assured that, just as the name suggests, the FNE really does last for just one night. Thereafter, it is sleep as usual, jet lag permitting.

## *Sleep deprivation*

New parents are renowned for stoically enduring a chronic shortage of sleep for the first few months or even years after welcoming their new arrival. Anyone suffering from such extreme sleep deprivation must feel like a contestant on *Shattered*, a 2004 UK reality-TV series shown on Channel 4 in which a hapless group of ten contestants fought it out to see who could stay awake the longest in order to win the £100,000 prize money.\* After all, when American DJ Peter Tripp tried this back in 1959,

\* *The Guinness book of records* stopped certifying attempts to break this particular record for fear of encouraging dangerous behaviour. This is rather ironic when you consider that they still certify the record for the longest tightrope walk over an active volcano crater!

he ended up suffering from a nervous breakdown after staying awake for a world record 201 hours – that is, more than eight days. The winner of *Shattered*, a nineteen-year-old police cadet by the name of Clare Southern, managed to claim the prize after keeping her eyes open for 178 hours.[31]

According to the sleep-deprivation literature, new parents are likely to be suffering from a lack of empathy and their ability to engage in social interactions will be severely compromised too. As if that was not bad enough, they will tend to be extremely irritable, impatient, unable to concentrate and permanently tired.[32] The good news here, though, for any new parents wanting to engage in a bit of sensehacking, is that many of the multisensory manipulations that have been shown to help adults get a decent night's sleep work even better for those at either end of the age spectrum.

For example, having a regular bedtime ritual of massage together with a bath scented with lavender not only helps to improve the quality and duration of sleep amongst infants and toddlers but improves maternal mood too. Stress levels in both infant and parent are also reduced. Developing effective sense hacks in this area is especially important given that one of the most frequent concerns of parents with young children relates to sleep problems, occurring in 20–30 per cent of cases.[33] I myself worked for a number of years as a sensory spokesperson for Johnson & Johnson, travelling the world speaking to paediatricians and nurses, trying to impress upon them the importance of a balanced diet of multisensory stimulation between neonates and their caregivers for better sleep.[34]

At the same time, however, mounting evidence suggests that school kids are simply not getting enough sleep. Indeed, according to the National Sleep Foundation, almost 90 per cent of high-school students do not get the recommended amount of sleep and, worse still, they are getting less year-on-year. It has been suggested that this poses a serious threat to health and academic success, and, more worryingly, the consequences may be irreversible. One tried-and-tested technique to improve academic performance in this case is simply to push back school start times.[35]

However, for those chronically sleep-deprived teenagers whose school opening hours have yet to change, one solution that has been shown to help them to get forty-three minutes more sleep a night involves presenting a brief sequence of camera-like flashes of light every twenty seconds (a bit like a very slow strobe light) for the last two or three hours of sleep, combined with a bit of cognitive behavioural therapy. According to the results of a randomized clinical trial from Stanford University, the treatment effectively helped reset the kids' body clocks without waking them, with the positive effects seen within a month of the trial starting. What is more, subsequent research from this group has shown that flashing a light every eight seconds doubled the shifting of the circadian clock compared to the twenty-second rate used in their original study.[36] This, then, might well be the best solution until more school boards take the step of delaying start times.

At the other end of the age spectrum, the scent of lavender has been shown to help the elderly, including agitated dementia patients.[37] Indeed, effectively sensehacking sleep is especially important for those who are getting on in years, as all the evidence points to the fact that the older you are, the less of it you get. According to one review, 50 per cent of older adults complain about the difficulty they have initiating or maintaining sleep. What is especially worrying is the way in which many older patients are kept on sleeping pills for long periods of time, even though they are only intended for short-term use.

## The scent of sleep

One study reported on an intervention carried out with four elderly psychogeriatric patients, three of whom had been on sleeping tablets for seven months, one year and three years, respectively.* After baseline testing for two weeks, the patients were taken off their hypnotic medication for a further two weeks. As might have been expected, they slept for an average of one less hour (this is known as rebound insomnia). The amazing result, however, was what happened in the final fortnight of the study, when the scent of lavender was diffused into the ward at night. Sleep levels returned to the same level that had been seen while the patients were on the hypnotic medication. This striking result led the authors behind the study to ask whether exposure to odour, sensehacking smell in other words, could potentially be more economical, not to mention safer, than current medication in treating sleeping problems amongst the elderly and infirm.[38] On the basis of their systematic review of the evidence published up until 2012, Fismer and Pilkington were led to suggest 'cautious optimism' regarding the beneficial effect of inhaling lavender. As seemingly is always the case with meta-analyses, though, the authors' recommendation was that more research was needed in order to support unequivocally lavender's effectiveness in helping people to relax and get a good night's sleep. That we are not yet able to say anything more concrete is perhaps all the more surprising, given that lavender's use as an aid to relaxation and sleep has appeared in plays and novels going back several centuries.[39]

Future research will in all likelihood demonstrate not only a pharmacological effect of the principal compound in lavender,

---

* However, note the very small sample size.

linalool, but also a psychological effect associated with the scent and where we may have experienced it previously. Supporting the former account, in one study heavily caffeinated mice were 92 per cent less active when in the presence of vaporized lavender oil. Intriguingly, though, these effects were eliminated in those mice that were unable to smell.[40] One of the problems when trying to evaluate aromatherapy claims such as the purported link between lavender and sleep/relaxation is that the researchers involved rarely specify which of the almost 500 different species of lavender they tested. What is more, it is not even clear whether there is any meaningful difference between synthetic odours and the natural essential oils.

## Sensehacking our dreams

Are you a lucid dreamer? This is the name given to those who seem to have an extraordinary ability to control their dreams. Lucid dreamers are aware that they are dreaming and are able to influence what happens while they are asleep. Evidence suggests that most of us will have a lucid dream at some point in our lives. However, there is a growing community of enthusiasts interested in trying to sensehack their dreams in order to dream lucidly more frequently. Could releasing scent, for example, influence our dreams? Over the years, researchers have tried everything from splashing water on people while they sleep through to the administration of light, sound, vibration and even movement (with the participants sleeping in a hammock) in order to try to influence dream content. However, while a burgeoning industry of technologies and techniques promises to augment lucid dreaming, the scientific evidence is pretty shaky at best. What is more, many of those conducting the research supporting the apparent effectiveness of these sensory interventions commercialize their insights,* thus leading to an inevitable concern over a perceived conflict of interest. Not that there is necessarily anything unethical going on, but I for one would prefer to see some properly controlled independent peer-reviewed research before splashing out on any of the proposed solutions.[41]

## Rise and shine: is it time to wake up and smell the bacon?

Do you dread the harsh and insistent sound of the alarm clock first thing in the morning? If so, you are not alone. Surveys often reveal that an alarm constitutes one of the least pleasant sounds that we are exposed to on a regular basis. Surely we could do better? In the olden days, it might well have been the factory whistle that would have been used to awaken the workers, and before that, the short-wavelength blue light of dawn. In the springtime in the UK, the dawn chorus, something that we hear less and less of these days, would once have provided nature's insistent wake-up call. For a while too, there were tea-making alarm clocks, remember them? The theory was that being woken up by a pleasant scent might be a little less jarring than being awakened by a sudden loud sound. These days, one can find a range of futuristic alarms that will artificially simulate the bright light of dawn.[42]

The mobile devices that most young people now use as a timepiece allow for a host of different wake-up options.† In

---

* Selling products with names like the DreamLight, the DreamLink, NovaDreamer, and the intriguing-sounding Hearne's Electric 'Dream Machine'.

† By 2011, almost 60 per cent of sixteen- to thirty-four-year-olds were using their phone as their primary means of keeping time (see https://today.yougov.com/topics/lifestyle/articles-reports/2011/05/05/brother-do-you-have-time). The figure is undoubtedly much higher today.

2014 Oscar Mayer produced a limited-edition bacon-scented wake-up app for the iPhone. People enthusiastically latched onto this meaty scent-enabled smartphone solution that played a sizzling sound while squirting out the scent of frying bacon from a plug-in capsule. This olfactorily enhanced multisensory intervention, mailed out to almost 5,000 people, was a huge marketing success, garnering a great deal of coverage in the media and online. That said, it is important to note that the smell by itself is unlikely to have helped you to wake up. This is contrary to the claims accompanying this marketing activation suggesting, 'When imagination blossoms, only this scent will guide you to your greatest awakening.'[43]

Worryingly for parents, it is not just smells that fail to wake us up. Many traditional smoke alarms simply will not wake up a child who is deep in slow-wave sleep. Personalized smoke alarms have been shown to work much better. For example, a pre-recording of one of the child's parents repeatedly addressing them by name followed by the instruction 'Wake up! Get out of bed! Leave the room!' roused all but one of twenty-four children in one study as compared to just over half when a traditional tone alarm sounded. Furthermore, more than twice as many children successfully performed the escape procedure, doing so, on average, in just twenty seconds as compared to three minutes for the traditional tone warning. If there really were to be a fire, such a simple sense hack could mean the difference between life and death.[44]

## Sleep inertia

It is not just falling asleep that so many of us find hard to do. Waking up can be difficult too. And then, as if that wasn't bad enough, there are the long-lasting after-effects of sleep known as 'sleep inertia', or what one researcher, writing in *Science* in 1968, engagingly labelled 'sleep drunkenness'. The duration and severity of cognitive impairment depend both on your recent sleep history and the stage of sleep that you wake up from. As might be expected, sleep inertia is more pronounced if you are awoken from deep, slow-wave sleep. According to one commentator, writing in the *Guardian* about an app designed to wake you from the lightest phase of sleep, 'The result is so gentle and lovely it feels like being woken up by a mermaid stroking your hair.' One group of researchers have suggested that the deleterious effects of sleep inertia on our cognitive performance can be felt for as long as two to four hours after waking, even after a full eight hours of sleep.[45]

No wonder, then, that so many of us reach for the coffee to help counteract this everyday lack of alertness. And for those who, for whatever reason, prefer to avoid caffeine, the good news is that even decaffeinated coffee can still speed up your reactions, at least if you happen to be a fan of the caffeinated stuff. Returning to the toothpaste example from earlier, it would be interesting to know whether filling our mouths with minty bubbles when we wash our teeth first thing in the morning has any effect on counteracting sleep inertia. My guess is that it does. Or perhaps better still in this regard would be the caffeinated Power Energy toothpaste launched by an enterprising start-up a few years ago.* The Colgate-Palmolive Company also filed a patent in 2013 for a toothbrush with a built-in caffeine patch. Taking a shower with a highly fragranced washing product presumably does much the same thing in terms of alerting us, though, as yet, there isn't so much research on this particular 'scent-sory' strategy.[46]

---

* Caffeine is absorbed more rapidly through the gums than through the lining of the stomach, though the effects are shorter lasting.

Sleep inertia is an especially pertinent problem for astronauts and long-haul airline pilots. Both groups of highly trained individuals occasionally need to wake suddenly in order to respond to unexpected in-flight emergencies. Indeed, it has been suggested that the 2010 Air India Express crash that resulted in more than 150 fatalities may have been caused in part by the poor decisions that were made by the captain who had just woken up from an in-flight nap. A lack of sleep has also been implicated in the Three Mile Island and Chernobyl nuclear disasters as well as in the *Exxon Valdez* oil spill (though alcohol may have played a part here too), not to mention the *Challenger* Space Shuttle disaster. The decisions made by on-call doctors and nurses who have just woken on the night shift are presumably also similarly adversely affected.[47]

Other than a bracing dose of coffee, one of the other sense hacks that may help to counteract the effects of sleep inertia involves not caffeine but melodic music. In an article published in 2020, a group of Australian researchers from Melbourne suggested that waking up to such music helps counteract the deficit in alertness that we all suffer when we wake up. They assessed the severity of the sleep inertia experienced by a group of fifty people as a function of the music that they started their day with using a self-report online questionnaire. Melodic music was the clear winner, at least when compared to being woken up by the sound of a regular alarm clock, the suggestion being that rhythmical music keeps our attention focused. And should you be wondering what the researchers had in mind then you need look no further than The Beach Boys' 'Good Vibrations' from 1966, or The Cure's 'Close to Me' from 1985. For those of you who prefer to start your day with something a little more classical, why not go for Beethoven's 'Für Elise' or Antonio Vivaldi's *The Four Seasons*.[48]

## Can dawn light wake you up?

Separately, there is a fairly extensive literature demonstrating the beneficial effects of exposure to bright light in the morning on subjective well-being, mood and cognitive performance. One study compared a bright artificial light source to a dawn-simulating light and a monochromatic blue light. While the benefits of bright artificial dawn light on cognitive performance were apparent only on the first night when sleep was restricted to just six hours, subjective mood and well-being ratings were higher following both nights in the lab. Exposure to the monochromatic blue light, meanwhile, showed signs of phase resetting of the circadian clock. Some forward-thinking airports use blue lighting to welcome passengers deplaning after a long-haul flight. And finally here, for those of you who have heard about the idea that shining a bright light behind your knees can be used to reset your circadian clock, I am afraid that the evidence does not support that assertion. So, my advice would be to keep your trousers on for now.[49]

The cult of early rising and short sleeping is finally starting to wane, as the profound benefits for our social, emotional and physical well-being of regularly getting a good night's sleep become increasingly apparent. Hence, figuring out the best way to enable more of us to spend a greater amount of our time asleep, while all the while maximizing our sleep efficiency, will become increasingly important. My best guess is that, in the years to come, sensehacking our sleep will be combined with cognitive behavioural therapy strategies. Tying multisensory interventions to the various stages of sleep using the latest in mobile technology can't be too far off either. Indeed, there are already sleep apps and mobile applications such as SleepBot, Somnuva, Zeez and Simba Sleep that promise to do just that.

Another option favoured by the (formerly) sleepless of Silicon Valley is the Oura Smart Ring. It tracks various physiological measures while the associated app makes recommendations for when you should go to sleep. No wonder that the globetrotting British royal Prince Harry was seen wearing one of these on his 2018 Australian tour. Twitter founder and CEO Jack Dorsey is also a fan. Such solutions should soon allow all of us, should we so wish, to optimize and/or personalize our own sleep hygiene,* no matter whether we happen to be an owl or a lark.

## *Are we really more sleep deprived than ever?*

Before leaving the subject of sleep, it is worth coming back to the question of whether we really are more sleep deprived than ever. This is undoubtedly the narrative heard from the press and many sleep researchers. The suggestion that we are all chronically sleep deprived, of course, also helps support the burgeoning sleep-hacking industry. It is clear that many people are worried about sleep, and it is equally clear that insufficient sleep is terrible for our health and well-being. Nevertheless, the pessimistic view that the majority of us are more sleep deprived than ever before actually goes against the broad position outlined by Steven Pinker and others that, on most metrics, life is much better today than it has ever been, despite what so many naysayers would have us believe. Look back in history and there is no shortage of famous people who complained of poor sleep, from Aristotle to Napoleon Bonaparte to Charles Dickens.

The key question is whether sleep is one of the areas that bucks the general positive trend. It may not be, at least according to the

---

* According to a 2019 survey from Lenus Health, something like one in four Brits already track their sleep.

latest survey study from my colleagues at Oxford University based on a careful analysis of time-use sleep diaries completed by a little over 18,000 UK residents at one of three time-points: 1974/75, 2000/01 and 2014/15. Their results suggest that we are actually getting an average of forty-five minutes *more* sleep a night than we did in the 1970s. Perhaps, then, the situation isn't quite as bad as some would have us believe. What is more, the latest results of a pre-registered cohort study also suggest that screen time at night has only a negligible effect on how much sleep our children are getting.[50] This is not to say that there aren't a number of us who wouldn't benefit from getting a little more shuteye than we currently do, but rather to remind ourselves that we may actually have never had it so good compared to those who slept before us. So let's just put our phones and tablets away and enjoy sleep a bit more.

# 5. Commuting

Let's be clear about this, driving is one of the most dangerous things we do. As I tell the female undergraduates in my psychology class, their most likely cause of death while studying at university is at the hands of their boyfriend while he is at the wheel of a car.[1] At the same time, however, driving is also one of the few activities that hasn't got faster as the decades have gone by.[2] Indeed, average speeds in the world's growing number of megacities (defined as those cities with more than 10 million inhabitants) is currently around nine miles an hour, and is predicted to drop to around two miles an hour by 2030. You would, in other words, often be better off walking or cycling. No wonder, then, that so many of those who commute by car find it so stressful.[3] North Americans spend an average of an hour a day behind the wheel.[4] So, driving that new car on the open roads with no other vehicle in sight – as the marketers so often like to show us – is contrary to the truth of the matter: we are all much more likely to be frustrated and stressed, stuck in a jam somewhere, inhaling traffic fumes.

After our homes, our cars are the single most expensive purchase that many of us will ever make. No wonder, then, that the manufacturers have got sensehacking down to a fine art. The visual appearance, the sound, the smell, even the feel, have all been carefully crafted to deliver just the right impression. That said, commuting is also in a state of flux, what with the changes in public transport usage brought about by Covid-19, the rise of electric and hybrid cars, semi-autonomous vehicles, and even driverless cars just over the horizon. Who knows, before too long, perhaps there will even be flying cars. But while the technological challenges of realizing these ideas are rapidly being surmounted, one should never forget the fundamental psychological challenges around what is, let's face it, a very unnatural activity. Our brains certainly did not evolve to drive, which is probably why so many of us get car sick.

## *What's real and what's not?*

The car is the paradigm case of multisensory design where everything has been engineered to communicate subliminally, through sight, sound, scent and even touch, just the right feeling to the driver. Everything from the sound of the car's engine through to the reassuring thud of the car door as it is closed, and from the weight of the car keys in the driver's hand through to that wonderful new-car smell, is carefully designed. Since the end of the Second World War there has probably been more research into hacking the driver's senses to provide the optimal multisensory experience while at the wheel than in any other environment. In no other field has sensehacking been turned into such a fine art, or, rather, science. In fact, careful consideration of how the multisensory experience of driving has been hacked will give you a pretty good insight into how those hacks might be, and in many cases have been, applied in other sectors, such as the 'smiley faces' on cars and other objects illustrated in Chapter 1.*

Let's start, though, with 'new-car smell'. The distinctive scent given off by a new car is probably one of the most positively valenced odours in the world. This is rather ironic, as the 'natural'

---

* Much of my academic career has involved taking insights and innovations from the world of automobile research and applying them to the design of everything from crisps to deodorant cans.

scent of a vehicle is a really rather unpleasant hint of fishiness from all the volatile organic compounds given off by the plastic interior as it swelters under the sun. Unless you are rich, long gone are the days of the authentic walnut trim and real leather upholstery. Nowadays, 'new-car smell' is almost always an artificial concoction from a fragrance lab. Even when you do find leather, which is more often a thin veneer rather than the real thing, it will have had the synthetic aroma of cowhide infused into it. Getting the interior smell right is something that the car companies take very seriously. There are whole teams of people working in car plants whose job it is to make sure that the cabin is impregnated with just the right mix of chemicals in order to deliver that distinctively pleasurable and rewarding smell when the driver takes delivery. Bizarre though it may sound, there are even annual lists of the best-smelling new cars.

There is, though, nothing fundamentally pleasurable about the scent of a new car. None of us, I think, was born liking it. Rather, we learn to like those smells that are associated with reward, be it gustatory in the case of foods, or high-value items as in the case of the car. The positive valence that so many of us attach to the smell of a new car is just a very powerful example of associative learning. Get it right, though, and you can really transform the experience. My all-time favourite example of this comes from the anecdotal reports of Rolls-Royce owners in the UK who would send their pride and joy off to the Midlands for a service or repair. They would get it back, and say, *'Wow! It is like new!'* Sure, it had had a good tune-up, and probably a polish, but the key change was the new-car smell, an aromatic mixture of leather and wood designed to capture the distinctive scent of a vintage 1965 Silver Cloud. This car cologne, if you will, is sprayed into the cabin just before the car is returned to the customer. According to Hugh Hadland, Managing Director of S. C.

Gordon Ltd, the coachbuilders of Rolls-Royce cars, 'People say they don't understand what we've done, but that their cars come back different and better.' So, why not take a tip from the experts when next you come to sell your own car. Give it a squirt of new-car scent, and you'll be sure to make your motor a more enticing proposition for whoever is taking it out for a test drive. It shouldn't make a difference, but all the research suggests that it most definitely does.[5] Hacking the senses, one smell at a time.

## *Vroom, vroom: just how important is the sound of the engine?*

Car manufacturers understand only too well that the owners of quality marques expect their car's engine to sound distinctively different from those of competitor brands. A Mercedes really does have to sound different from a BMW, say, or a Porsche. That said, engineering has developed to such a point that the interior of the car can be sonically isolated from pretty much everything else that is going on outside. What this means in practice is that the vehicle can be made effectively silent to those inside. However, this is very definitely not what the driver wants (not) to hear, nor what they have paid all that money for. They want to be made aware of the distinctive sound of the engine, convincing them of what a good purchase they have made with every roar. No wonder, then, that psychoacousticians have put a lot of work into identifying the particular qualities that the engine sound should have.[6] What this means, in practice, is that after having successfully designed all of the noise out of the car's interior, the engineers spend almost as much time working out how to put it back in. It is just that the sound of the engine you hear is most likely artificial. Some, in fact, have compared it to lip-syncing.

In 2015 General Motors submitted patents for an electronic

method of generating engine noise, while some models of the Volkswagen Golf have 'soundaktors' installed. These sound actuators help boost the roar of the engine. The engineers have now started adding a growl to some family cars too in order to convince drivers that their vehicle is just that little bit more powerful.[7] For instance, the Peugeot 308 GTi has a number of different driving settings, and should the driver select 'sport' mode then the engine will suddenly start to growl. Not only that, but the background lighting inside the car turns from white to bright red. Does something so simple really convince people that the car that they are driving is more powerful? Well, according to the research, this kind of approach just might work.

Laboratory research involving students playing video driving games suggests that louder cars appear to be travelling faster.[8] Meanwhile, in another study, lowering the level of realistic in-car noise by five decibels led those who were watching driving videos to underestimate an actual travel speed of 60 kilometres per hour by 10 per cent.[9] This could be because drivers treat engine noise as a diagnostic cue concerning the speed of the vehicle. Alternatively, the sound may directly affect how visual cues to speed, such as the rate at which the scenery passes by, are perceived as the result of multisensory integration. As for the red lighting, the research shows that both red cars and red trains do indeed sound louder to those who can see them than those of an other colour.[10]

The growing realization amongst the public that the engine sounds we hear might not be 'real' has come to the fore recently as debate concerning the dangers posed by electric cars continues to rage. While silent cars might seem like a good idea, especially given how many complaints there are about road traffic noise, it is important to remember that silence can also be deadly. At low speeds, electric cars are effectively silent and, hence, pedestrians and other road users, including the visually impaired, simply do not know that they are there. Indeed, a 2018 report suggested that pedestrians were 40 per cent more likely to be hit by a hybrid or electric car than by a conventional diesel or petrol car.[11] No wonder, then, that many countries now require electric cars to make an artificial engine noise at low speeds. Precisely what the electric car should sound like, though, is just one of the questions that psychoacousticians and marketing agencies continue to grapple with.*

## Can you hear the quality?

The sound of the car door is a little different. Closing a car door was never going to be silent. Once again, though, as the advertisers know only too well, it is the solid *thunk* of security when a well-built door is shut that helps reassure the consumer in the showroom of what a good purchase they have in front of them. It is just the sort of sound that really can help to seal the deal. Car companies like VW and Renault have spent a lot of time engineering the sound of their car doors. And, having managed to establish what they considered to be a truly great sound, VW used it relentlessly in their 'Just like a Golf' TV adverts. The campaign was based on the notion that the distinctive sound of VW Golf car doors closing was simply much better than that of any other similar model. The claim implicit in these ads was that you could literally hear the quality.

However, beyond the sound of the car door, one of my other all-time favourite examples of sound design relates to the behaviour that a third of us apparently engage in when visiting the car

---

* This is reminiscent of the question that must once have faced the set designers of the original *Star Wars* movie, namely what should a lightsaber sound like?

showroom: tapping on the dashboard and listening to the sound that our knuckles make. This is something that no one in their right mind ever does once they have made a purchase. But at the moment when the customer is in the car showroom, trying to decide which model to buy, it can make all the difference.[12] No wonder, then, that some car companies also work to make sure that their dashboards sound just right when tapped. That, along with the sound of the horn, turns out to be a surprisingly important driver of car sales.[13]

At the luxury end of the market, nothing is left to chance. For instance, in the Bentley Continental GT, even the sound of the indicators has been carefully crafted to emulate the tick-tock of a carriage clock. This particular sound was chosen because of its association with history and heritage, culture and class.

## *'Make it snuggle in the palm'*

Touch is important here too. Just think about the weight of the car keys in your hand. They need to feel just right, don't they? Like the auditory and olfactory cues that we have just mentioned, touch can be used as a subtle sensory cue that can seal the sale. And it is here that I am reminded of the approach forwarded by Sheldon and Arens during the Great Depression. One of their key recommendations, if you'll excuse the pun, from their 1932 book titled *Consumer engineering* was, 'Make it snuggle in the palm.' These pioneering researchers are worth quoting at length given their thoughts on the importance of the feel of automobiles:

> After the eye, the hand is the first censor to pass on acceptance, and if the hand's judgment is unfavorable, the most attractive object will not gain the popularity it deserves. On the other

hand, merchandise designed to be pleasing to the hand wins an approval that may never register in the mind, but which will determine additional purchases. The deciding factor in the purchase of an automobile may be not free-wheeling, or chromium accessory gadgets, but the feel of the door handles, the steering wheel, upholstery.[14]

'You can have any colour as long as it's black.' This, in case you don't recognize it, is the famous line said to be from American car-maker and industrialist Henry Ford.* But the colour of your car matters, probably more than you think. According to one analysis of more than 2 million online car sales, it turns out that people simply aren't willing to pay anything like as much for a gold-coloured second-hand motor as for one of any other colour. Yellow cars, perhaps because of their rarity, appear to retain their value better (though, I presume, in some countries this colour may be associated with taxis more than anything else). Meanwhile, for those in trouble with the traffic cops, red is probably the last colour you want, given that people believe red cars are going faster, and seem to make more noise, than cars of any other colour.[15]

## What has techno got to do with RTAs?

However, it is not just the colour of the paint or the sound of the engine that influences how fast a car appears to be travelling. At least, not according to research in which students drove in a video game or simulator while listening to different styles of music. No prizes for guessing that those listening to high-paced techno

---

* This statement is not quite correct, as black is actually achromatic, defined by the very absence of colour. The provenance of the statement has also been questioned.

music drove faster and broke more virtual traffic laws than those listening to something a little calmer.[16] Intriguingly, though, it is not that drivers stop paying attention to the other traffic altogether under such highly arousing conditions. Rather, their visual attention becomes much more narrowly focused on the roadway directly in front of their vehicle, meaning that they tend to miss more of what is going on in the periphery.[17] Given such findings, we should perhaps all be worried by the results of one US study showing that many young male drivers listen to their car stereos at deafening levels of 83–130dB.[18] (Of course, it's also possible that those who like techno are likely to be the sort of driver who would speed anyway.)

In terms of hacking the driver's senses to help promote safer driving, the South Korean car manufacturer Hyundai suggested at the 2018 Geneva Motor Show that it was even considering piping relaxing music into some of their models in order to help combat road rage. If the car detects that a driver is stressed, it could search a music-streaming service such as Spotify for some 'soothing' music to play (or should that be to pacify), while at the same time, perhaps, dimming the interior lighting.[19]

## *Distracted by technology: capturing the attention of the distracted driver*

Talking on the phone while driving increases the risk of having an accident fourfold. This is roughly equivalent to what is seen as the drink/driving limit in many countries is reached.[20] Remarkably, the manual aspect of holding the phone isn't the main problem here. Hand-held or hands-free, the risks are roughly the same. Rather, it is the inability to divide attention effectively between eye and ear that is the fundamental issue. Particularly challenging in this regard is the fact that our auditory attention is focused on whoever we are listening to (e.g. on a mobile device), while our visual attention should be on the road ahead. The issue with so much of the modern technology that we have is that it allows different streams of information to be presented from different directions simultaneously.*

Dividing attention between different locations is something our brain finds it difficult to do. This was, in fact, the very problem with which I started my academic research career back in 1990.[21] In an undergraduate research project, I was able to show that people find it harder to listen to what someone else is saying under noisy conditions if the sound of the speaker's voice and the sight of their lips come from different locations. After all, we presumably evolved to pay attention to all of the sensory inputs coming from one location, integrating the sight and sound, be it of predator or prey, so that we could respond more rapidly. Unfortunately, however, in my experience, engineers rarely think about such cognitive constraints on our attentional resources when they are designing interfaces and warning signals.

In the early 2000s, I was lucky enough to be able to extend my undergraduate research from the psychology laboratory to a much more realistic setting, working with Dr Lily Read from the University of Leeds. We conducted a study in a high-fidelity driving simulator in which people had to drive a challenging route along a virtual road network while selectively repeating back one of two voices. Sometimes they had to repeat what the voice coming from directly in front of them was saying while ignoring the voice from the side. At other times, they had to do the opposite. The ability to dual task while driving was slightly,

---

* If you think this is no different from a driver talking to a passenger, you're wrong. A passenger is aware of the road conditions, so knows when the driver needs to concentrate more, and can adapt or pause the conversation accordingly.

but significantly, better when our participants were listening from the front rather than from the side.[22] The implications were clear. Having the voice of whomever we happen to be speaking to come from a talking windscreen ought to make it significantly easier for the driver of the future to pay attention to two things at once, and so, presumably, also make driving that little bit safer too. Sadly, the idea never took off. The downside was that, while it is undeniably easier to look and listen in the same direction, it is also much harder to block out the sound of someone else's voice when the road conditions become too demanding if their voice happens to come from the direction in which you are looking.*

### 'It can wait'

Texting at the wheel is the single biggest killer on our roads. People simply do not realize how long they take their eyes off the road when texting. It is a literally deadly activity. The research shows that your risk of having an accident while texting at the wheel goes up by a staggering twenty-three times.[23] Worse still, drivers have the false impression that they are aware of everything that is going on around them, even though in-car studies repeatedly show texting drivers looking away from the road for eight seconds or more. The scariest thing is that, when asked, drivers are convinced that they only looked away momentarily. Given such figures, you will hopefully understand why I am so frequently on the conference circuit highlighting the dangers and agitating for a change in the law. I am not the only one who is worried. Alarmed by the bad press when yet another texting

---

* Consider here how many drivers turn the radio down or off when faced with challenging driving conditions.

teen is maimed or killed at the wheel, some mobile companies have themselves been running hard-hitting campaigns to highlight the dangers too, such as AT&T's powerful 'It can wait' campaign in the USA. However, while texting at the wheel is still possible, perhaps the only way to counteract all the distracting technology in the car is to develop more effective warning signals to grab the driver's attention and direct it back to the road, no matter how exciting the last incoming message might be. To this end, I and my colleagues at the Crossmodal Research Laboratory in Oxford have spent the last couple of decades working with some of the world's largest car manufacturers on the design of enhanced warning signals to more effectively hack the driver's brain.[24]

We have, for instance, shown that multisensory alerts are more effective at capturing a distracted driver's attention than those that stimulate only one sense at a time. We have also demonstrated that multisensory warning signals work better if they are designed to imitate the kinds of stimuli that we believe our brains evolved to deal with, namely multiple sensory signals coming from the same direction at more or less the same time. Once again, this is not something that most engineers seem to have picked up on yet.

Just think about how much you jump if someone sneaks up and suddenly surprises you from behind. We were able to capture this phenomenon in a novel warning signal. We demonstrated that simply by presenting a sound from just behind the driver's head, e.g. via a loudspeaker mounted in the headrest, we could deliver a warning signal that was much more effective in terms of orienting their gaze back to the road than anything else that anyone had yet come up with. But why should any sound emanating from this particular region of space be so effective? It's because there are special circuits in our brain that monitor the space directly behind our heads. This is a space that we rarely see other than in a mirror, and hence rarely think about. A broad-frequency noise burst within 70cm (27.6in) of the back of the head, though, in a region that the cognitive neuroscientists refer to as near-rear peripersonal space, triggers an automatic and involuntary defensive response.[25] This is, then, an ideal place from which to present a warning signal if you want to hack the driver's brain and get their eyes back onto the road.

## Asleep at the wheel

Drowsiness while driving is another major issue on our roads today. To give you some sense of the scale of the problem, the results of one survey of 1,000 Australian drivers suggested that 80 per cent reported having driven while sleepy, with 20 per cent admitting to being relatively frequent 'drowsy drivers'.[26] Of all road traffic accidents, 10–30 per cent are attributable to the driver falling asleep at the wheel.[27] Under such conditions, one needs warning signals that will effectively wake the driver up. Back in the day, at a time when getting the population down into nuclear bunkers was the most pressing concern, there was some wonderful research funded by the US military intended to optimize the design of auditory alerts. In 1963 Oyer and Hardick published the results of their extensive research assessing the alerting potential of several hundred different sounds from foghorns to klaxons, pure tones to noise bursts. My personal favourites were the sounds of stampeding elephants and screaming babies. Neither, I hasten to add, was ever introduced.[28] However, while some alerting sounds were more effective than others, the problem is simply that attention capture, or alerting potential, tends to scale with unpleasantness.[29] The challenge is therefore to make alerting signals that wake the driver without any of the unpleasantness of a 100dB klaxon blasting out. Remember here that the driver is likely to hear this alerting sound more frequently than the general public would ever hear the sound indicating that they should immediately make their way to the nearest nuclear shelter. In our own research, we have worked extensively with the sound of a car horn as a semantically meaningful signal, one that is both intuitive and immediately recognizable. That sound effectively conveys the idea that there is something going on that demands attention.[30]

Others have talked about giving the driver an electric shock to wake them up.[31] That, or vibrating their buttocks if the intelligent transport system detects that the car is slowly deviating from the lane – this being suggestive of the driver having fallen asleep at the wheel.[32] These days, more cars than ever before are vibrating their drivers in order to convey some information or alert. And while it might sound crazy, just think about it for a moment. The skin is your biggest sensor, accounting as it does for 16–18 per cent of body mass, and yet it is rarely used while driving. Researchers have tried vibrating everything from the foot pedals to the back of your seat, and from the seatbelt to the steering wheel – any surface of the car, in fact, that the driver is in contact with.

In one of my all-time favourite studies, published back in 1967, John W. Senders put drivers on a test track with a visor that obscured their vision. When the visor came down, Senders would start his stopwatch, and measure how long they could continue driving before they needed to see the road again. He also tried attenuating the drivers' hearing with heavy-duty ear muffs, blocking their sense of touch using thick gloves and cutting off their sense of smell with a nose clip. No surprise that the amount of time we can keep driving blind is much shorter than we can manage following the removal of sound, touch or smell. In fact, this was the study that led to the statistic that one finds quoted everywhere in driving literature: that 90 per cent of driving is visual.[33]

And while this study may sound scary enough, Senders subsequently went on to repeat the design with Boston taxi drivers on the open road, to assess the role of expertise. This is the sort of study that could only have been done in the days before anyone had heard of Ethics Panels. I do, though, have to be careful what I say here, as poking fun at Senders's work once got me into hot water with the author's daughter. Unbeknownst to me, she was in the audience of a talk that I was giving at Microsoft in Cambridge. The old man himself was still going strong last time I heard from him, still interested in research as he approached his century, and we continued to exchange emails about the odd scientific question until shortly before his death in 2019.

## Driving the nature effect

However, rather than simply making warning signals to wake up the drowsy driver more effectively, one might also think more carefully about the scenery that they happen to be passing through. That was the intuition behind Highways England's 2018 announcement that they were going to invest £15 billion in order to improve a number of motorways and major A-roads by 2021. Their belief was that boring straight roads lead to drowsiness at the wheel. Introducing more scenic roadways should, or so the logic went, help to reduce the likelihood of drivers falling asleep at the wheel. Use the nature effect that we encountered earlier, in other words, to help make driving safer. But what exactly does exposure to nature do to us while we are driving? The research certainly suggests that driving through nature may be less detrimental to our mental performance than cruising through a semi-urban or built-up environment.

In one study, drivers spent longer trying to crack impossible anagrams, thus suggesting that they were less easily frustrated, after watching a short video of a drive along a scenic parkway than after watching an equivalent-length drive through a garden highway or built-up highway.[34] Meanwhile, in another study, drivers exhibited greater recovery from stress, as well as less anger, aggression and fear, when exposed to videos of vegetated rather than urban roadside scenes.[35] That said, one might worry that for those who are actually doing the driving, such scenery, no matter how pretty, will fade into the background while they keep their attention focused on the road ahead.[36] On the other hand, driving through the 'tunnelled' roads that one finds when road traffic authorities attempt to reduce traffic noise in built-up areas can be even worse. That so many of us find this a most unpleasant experience amply illustrates how important having some kind of view is to drivers.[37]

Long before there were cars, Renaissance architect Leon Battista Alberti wrote that roads should be made 'rich with pleasant scenery'.[38] True to this suggestion, in the late 1920s the Taconic State Parkway Commission built a 103-mile highway connecting New York City to several state parks and the Catskill and Adirondack Mountains. This road is unusual in that it was built primarily for the pleasure that it would give those who would one day drive along its winding route. Writing in *The New York Times*, journalist Mark Healy describes his journey along this particular parkway, running from Westchester to Columbia County, as 'like a dream', while another writer described it as 'a 110-mile-long postcard. It's the most beautiful road I've ever known – in all seasons.' 'There are no billboards on the Taconic,' Healy observes, 'no hideous rest stops, no tolls, no guard rails and no trucks. What there is, in abundance, is trees – trees lined up along the roadside and clusters of oaks, pines, and maples on the median.'[39] Given that it is hard to imagine anyone being crazy enough to pay for such a project nowadays, one can only applaud the foresight of those who originally came up with the idea, notably funded under the direction of Franklin D. Roosevelt. It stands out as a testament to the enduring appeal of nature, even when seen from behind the steering wheel. It would, though, be interesting to know a little more about how accident rates compare on these scenic roads to equivalent boring drives.

Changes to our driving behaviour are illustrated by the following quote from a 1938 volume entitled *American highways and roadsides*. According to its author, J. L. Gubbels:

> The best road between A and B is the road that is economical, safe and interesting. It has been estimated that sixty-five per cent of highway traffic is for pleasure, and pleasure comes from variety. What does the autoist care whether the measured distance from A to B is forty miles or forty-three miles? If the three extra miles will enable him to catch a long vista of a distant stream, see from a hilltop a broad valley where cattle, horses and sheep are grazing, see farmers mowing hay or turning the dark sod, or find his road ahead bending mysteriously under the arch of an overhanging tree, he will not begrudge the few minutes they cost him.[40]

Given that the urban environment not only looks different from the natural landscape but also sounds and smells different,* one might well wonder whether such auditory and olfactory cues influence our mood and state of mind, irrespective of any pleasure that we might get from driving. After all, it is much

---

* Though as Appleyard, Lynch and Myer wrote in 1965 (p. 17), 'the sensation of driving a car is primarily one of motion and space, felt in a continuous sequence. Vision, rather than sound or smell, is the principal sense ... Sounds, smells, sensations of touch and weather are all diluted in comparison with what the pedestrian experiences.'

more likely to be air pollution that irritates your nostrils when stuck in a traffic jam downtown than when winding along the Hudson Valley on the Taconic State Parkway. Consistent with the view that smell really does matter, researchers have highlighted how the number of road traffic accidents reported in LA correlates with the level of air pollution.[41] On reflection, perhaps this isn't all that surprising given that the more pollution there is, the more cars one might expect there to be on the roads causing it, everything else being equal.

On the flipside, however, with the windows up and the aircon on full blast, one might simply not get to smell nature, even when out driving through the fragrant countryside. This was part of the reason why, a few years ago, one British company decided to look into developing a scent display for their vehicles. The idea was for the GPS to periodically check on the car's location and instruct the device to pump the appropriate synthetic nature scent around the cabin. Just imagine how much more pleasant it would be to drive through the forest if your nostrils were stimulated by the scent of pine, say, or perhaps the wonderful smell of the earth just after rain. Sounds like an intriguing idea, multisensory congruency through and through.

If all this talk of scent displays sounds a little far-fetched to you, consider Citroën: a few years ago they launched their C4 model with a nine-scent display operating through the ventilation system. The scents were split into three groups of three, designed to be congruent with notions of 'travel', 'vitality' and 'well-being'.[42] The first three replacement cartridges were free, meaning that the car would be scented for six months before the driver needed to buy refills. In 2014 Mercedes offered its customers an olfactory display in some of its new models too.[43] My suspicion, though, is that it would end up being hard to convince people that it was worth their while paying for the refill.[44]

Of course, no one needs a high-tech gadget in order to deliver the synthetic scent of nature. Drivers have been dangling scent-infused cardboard pine trees from their rear-view mirrors for decades. However, the problem with that solution is that our brains tend to adapt pretty quickly to pleasant or neutral smells (as we saw in the Home chapter). So while you might notice the scent as you open your car door, my guess is that you probably won't think about it much after that.[45] It is for this very reason, then, that the periodic release of scent holds much more promise as far as the delivery of scenery-congruent scents is concerned, at least if you want drivers to pay attention to what they are smelling.

Pumping out peppermint aroma for a few seconds every few minutes improves people's cognitive performance on a range of repetitive, not to say rather boring, behavioural tasks.[46] (I'm allowed to say that, given that I was the co-author on one of the studies.) My best guess, therefore, is that scent displays will one day be used to arouse dozy drivers rather than the loud and unpleasant auditory alerts we came across earlier. Pumping out an arousing ambient scent such as cinnamon, peppermint, rosemary, eucalyptus or lemon might prove to be a much less aversive way of achieving the same result.

Stimulating two senses is normally better than stimulating one: enter the Japanese team who tried to hit the sleepy truck driver with everything they had got. This included an electrical finger massage delivered via the steering wheel, a burst of oxygen, a whiff of grapefruit scent, and the *pièce de résistance*, some shredded dried squid to chew on.* Nevertheless, verbal reports from the nine truckers who took part in this particular study suggested that the more sensory cues the better in terms of keeping them alert.[47] Any tired driver should, of course, ideally take a break. But, until such time as we can ensure that they actually do, a little olfactory

---

* No, I am not sure about the last one either.

pick-me-up (not forgetting the dried squid) might just prove to be the most effective means of hacking the driver's senses.

In a similar vein, others have been looking into the use of calming scents such as lavender to reduce the incidence of road rage.[48] Releasing an aromatherapy essential oil might well help stressed drivers to calm down and hence drive a little more safely.[49] And, of course, if it helps mask any of those unpleasant smells associated with air pollution, all the better.[50] Add in a little massage from the driver's seat and I am sure that everything will be right again in no time at all.

By monitoring their physiological signals, so-called intelligent transportation systems can now keep track of how stressed, relaxed or drowsy a driver is. Tell-tale signs of stress include clenching the steering wheel too tightly and/or braking too abruptly. Blinking too often, or characteristic changes in the tone of voice and in the patterns of intonation of any vocal responses that the driver makes can also provide useful clues as to a driver's state of mind. However, it is not just the driver's voice that is relevant here. It is possible to change the likelihood of a driver complying with an in-car verbal command simply by changing the voice. Barking out instructions like the military drill instructor in *Full Metal Jacket* will probably elicit a very different (i.e. more prompt) response than if the driver assistant has the silky tones of an Alexa or Siri.[51]

## *A risky solution*

Ultimately, no matter what safety interventions are introduced, the real challenge comes from the phenomenon of risk compensation. The evidence shows that we tend to adjust our driving behaviour so as to achieve a certain level of perceived risk. Hence, just as soon as some new intervention or other comes along to make driving safer – the mandatory use of seatbelts, or anti-lock brakes, for example – then drivers start to take more risks, safe in the knowledge that their car will (presumably) protect them, no matter what they get up to on the road.[52] One of my favourite suggestions is the counterintuitive idea of putting a spike, be it real or virtual, on the steering wheel.[53] This would dramatically increase the perceived risk for the driver, and so make them drive more cautiously as a result. Genius, no?

To be honest, placing a metal spike on the steering wheel isn't practicable, but is there anything else we could do to nudge the balance in terms of a driver's risk perception? I was engaged in just such a project with a global car manufacturer a few years ago. Our goal was to hack the driver's brain into feeling a little more fearful without their realizing quite what was going on. The idea was simple enough: we would project scary images subliminally onto the driver's side of the windscreen. After all, the cognitive neuroscience research shows that when scary faces – for example, seeing the whites of someone's eyes* – are presented, even if they are flashed up too briefly for anyone to register, they can nevertheless still activate the brain's fear circuits.[54] Sadly, though, as sometimes happens, we were never able to quite make this particular sense hack work in practice. It did, though, get us some great, if bemused, coverage when the international press heard about the scheme.

## *'Scared sick': why exactly do we get car sick?*

While a car is in steady motion, our body tells us that we are not moving, yet our other senses tell us that we are. Our sense of

---

* Or better still, just imagine Jack Nicholson's face when playing the deranged caretaker in the 'Here's Johnny!' scene from *The Shining*.

body position and movement, proprioception and kinaesthesia, respectively, informs our brain that we are static, whereas the liquid sloshing around in the three semicircular canals of the inner ear, what is known as the vestibular sense, tells our brain that we really *are* moving. Sensory incongruence, or conflict, also happens to be one of the consequences of having ingested certain neurotoxins. Hence, from an evolutionary perspective, it makes sense to try to eliminate the potential poison by vomiting out whatever you ate last.[55] That said, in terms of avoiding car sickness in the first place, the best thing to do is keep your eyes on the passing scenery and hope that vision will do what it normally does and dominate over the discrepant inputs from the other senses, convincing *all* of them that you really are moving, even if it doesn't necessarily feel like that.\*

The fact that our brains did not evolve with driving in mind may help to explain why it is that so many people suffer from car sickness, especially when, as a passenger, they try to concentrate on something other than the road, for example, reading a book. After all, vomiting out our last nutritious meal would not seem like a particularly sensible behaviour, evolutionarily speaking. One intriguing suggestion to try to explain this apparently maladaptive behaviour comes from Professor Michel Treisman, whose lab I took over when I started teaching in Oxford back in

---

\* While I myself normally do not get car sick, I do remember all too well the time when I was a young child of ten or so, sitting in the back of a windowless van facing backwards on a trip from Leeds to Ilkley, a journey of about eleven miles. Oh boy, was I ill when we arrived at our destination! In this case, though, I think it was the unusual patterns of acceleration and deceleration (given I was facing backwards) that did for me, paired with the fact that I had no visual cues concerning movement. I can assure you that once was enough! I certainly never made that mistake again. Illness-induced learning is amongst the most rapid and robust responses we exhibit.

1997.* In an article that appeared in the well-respected journal *Science*, Michel speculated that it might be the mismatch between the inputs coming to our different senses that is the problem for the 25–50 per cent of us who are affected.

For those of you who are feeling smug about not suffering from car sickness, just you wait. The problem is likely to hit you with a vengeance just as soon as we start driving around in semi-autonomous vehicles. Imagine the situation: there you are, waiting for the rare occasion when your vehicle tells you that you need to take back control. This creates a real challenge for cognitive ergonomists; drivers are likely to find sitting in a state of permanent readiness, while doing absolutely nothing most of the time, to be extremely boring.[56] However, if the driver is entertained, perhaps watching a favourite movie or Netflix show, then this will likely help the time pass while at the same time ensuring that the driver is alert enough to take over control suddenly should the need arise. However, this is also likely to create the ideal conditions to induce car sickness – namely, a person concentrating on something other than the movement of the vehicle in which they are travelling. All is not lost, though, as patents have now been filed by scientists claiming to have come up with a sensehacking solution that effectively tackles the problem by, once again, tricking the brain by means of intelligent glasses that beam light into occupants' peripheral vision to mimic movements on the outside of the vehicle. Figuring out an effective solution to this particular problem is likely going to be crucial, given that the market for driverless cars is expected to be worth £63 billion to the global economy by 2035.[57]

---

* And, for those who can still remember, no, he wasn't one of the professors that I mentioned in the first chapter.

## The road ahead

If there is one thing that is sure, it is that travel, and that includes commuting, is set for some radical disruption in the years ahead. Electric vehicles are already with us, with Tesla launching a number of new models, including its cheapest offering yet, the Model 3, in 2017. Semi-autonomous driving is also now legal in a number of states in North America and similar schemes look set to be rolled out elsewhere around the globe in the very near future (fatal accidents notwithstanding). No wonder, then, that with so much change on the cards, even the largest car manufacturers are wondering whether there will still be a future for them producing conventional vehicles in the decades to come. These days, disruption to the personal transportation industry is much more likely to come from the likes of Google or Apple, nuTonomy, Lyft or Uber than it is from industry stalwarts like Ford or Toyota.[58] Be it the rise of electrification, car sharing or autonomous driving, the traditional brands that car companies have spent so much money building up over the years with expensive marketing campaigns are looking increasingly vulnerable. They may well struggle to stay relevant if car transport increasingly comes to resemble the ride-hailing app sector.[59] But whatever happens, it is clear that the most successful players in this sector will need to figure out how best to sensehack the commuter's brain, given it didn't evolve to drive to work, or anywhere else either, for that matter.

# 6. Workplace

The Japanese have a word for those who literally work themselves to death – *karoshi*. So serious is the problem that in 2018 the government there was forced to introduce legislation limiting the amount of overtime that workers are allowed to perform – no more than 100 hours in a single month, and a maximum of 720 hours in a year. While the culture of such long office hours is perhaps more common in Japan than elsewhere, workers across the globe spend more of their waking lives indoors than pretty much anywhere else. In the US, for example, the average working week currently comes in at a little over 34 hours. This compares to a maximum average of almost 43 hours a week in Mexico and a minimum average of just 26.5 hours in Germany. Meanwhile, according to a 2019 press report, Brits put in the longest office hours in Europe, with newspaper headlines suggesting 'Long hours at the office can leave you "jetlagged".'[1] These national averages undoubtedly hide a great deal of individual variation, with many people reporting that they regularly work 60–70-hour weeks. In fact, according to a report in the *Harvard Business Review*, 62 per cent of high-earning individuals work more than 50 hours a week, 35 per cent work more than 60 hours a week and 10 per cent work more than 80 hours.[2] One of the ironies here, at least according to John Pencavel from Stanford University, is that analysis of female munitions workers pulling 70-hour weeks during the First World War (this a classic study in the field) revealed that they did not get much more done than those working just 56 hours.[3]

Spending so much time working would not be such a bad thing if more people actually enjoyed it, but this is often not the case. In fact, survey after survey highlights that stress levels and disengagement amongst workers are running at record highs. For instance, according to a 2011–12 Gallup investigation, the cost to companies in terms of lost productivity from distracted and disengaged workers was somewhere in the region of $450–550 billion a year in the US alone. According to the report's authors, 'By the end of 2012, . . . only 30% of American workers were engaged, involved in, enthusiastic about, and committed to their workplace.' Of the remainder, 52 per cent were not engaged while the remaining 18 per cent admitted to being actively disengaged. Meanwhile, more than 70 per cent of the Australians in a 2017 survey reported that work made them stressed.[4]

Work-related stress plays a key role in many non-communicable diseases in urban societies today, including everything from cardiovascular disease to depression, and from musculoskeletal diseases to back pain. While some trendy offices have attempted to tackle the problem by introducing everything from slides and rock-climbing walls (see, for example, the glass climbing wall at 22 Bishopsgate, a skyscraper in the City of London), through to go-karts and even shooting ranges, the word from many workers is that they actually prefer support and recognition.[5] Other solutions here include massage, with Tiffany Field and her colleagues in Florida showing that a fifteen-minute massage at lunchtime enhances people's concentration in the afternoon. The lucky participants taking part in this study worked in medical research and were given a daily massage for five weeks. Sounds great, doesn't it?[6]

## *Sensory imbalance in the workplace*

When you consider that we did not evolve to spend 90 per cent of our lives indoors, these negative outcomes should perhaps not come as such a surprise. Spending so much time indoors (i.e. where most of us work) has been shown to lead to a host of health problems linked to an imbalance of sensory stimulation. For instance, in northern latitudes, seasonal affective disorder (SAD) is a major problem. The chronic shortage of natural light in the winter months can all too easily depress those who rarely make it outside during the all-too-brief daylight office hours. To give some sense of the scale of the problem, it has been estimated that as many as 2 million workers in Manhattan alone may suffer from the negative consequences of light hunger during the winter months. Fortunately, however, the solution in this case is simply to increase one's exposure to artificial bright light that mimics natural daylight – that, or emigrate somewhere sunny and warm. In fact, ensuring adequate lighting turns out to be one of the simplest and most effective sense hacks enhancing our performance and well-being at work.[7]

## *Anyone for lean design?*

The very nature of the workplace has changed as the years have gone by. For instance, Josiah Wedgwood, the eighteenth-century British industrialist, has been credited with the idea of keeping the workplace clean. He is probably the one you should blame for the now widespread 'lean' approach to office design.[8] As noted by *The Economist* in 2019, however, office design is constantly changing:

At the start of the twentieth century offices aimed to maximize efficiency by mimicking the factory layout with rows of supervised typists and clerks, as promoted by Frederick Taylor, an early American management consultant. In the 1960s, less rigid *Bürolandschaft* ('office landscaping') made its way across the Channel from Germany. The 1980s ushered in 'cubicle farms'. Today open-plan offices and unassigned 'hot desks' aim to flatten hierarchies and increase informality.[9]

## Talk about a sick building

Just to add to the list of acronyms, here is another one for you, 'SBS', standing for 'sick building syndrome'. This is the name given to those buildings where complaints of ill-health are more common than expected. In 1982, the World Health Organization defined the condition as 'a set of general, mucosal and skin symptoms perceived by the inhabitants of buildings with indoor climate problems'. The most common symptoms are lethargy, headaches, nose and/or throat irritation and eye irritation. According to estimates from Sweden, around 12 per cent of female office workers and 4 per cent of males are affected. At the turn of the century, the cost to the UK economy was estimated at around £600 million a year, amounting to something like 2 per cent of a company's payroll.

The first reports of SBS started to appear in the West during the oil embargo of the 1970s, when ventilation standards in many office buildings were lowered. Indeed, SBS tends to be more common in predominantly sealed offices with little natural ventilation than in those buildings where the staff can open the windows to get some fresh air. A lack of proper ventilation can all too easily lead to the build-up of volatile organic compounds (VOCs) emitted by the furniture and coatings in office buildings.

Carbon dioxide levels may also be higher than ideal in such buildings because everyone breathes this gas out continuously. Many outbreaks of SBS have been linked to air pollution and/or a 'strange' (i.e. unfamiliar) smell in the air – though most descriptions don't get any more specific than that in terms of what smells you should be looking out for (if that is the right expression).[10]

Reports of SBS in the workplace do seem to have declined somewhat in recent years. Furthermore, the inability to establish a clear causal mechanism has led some commentators to wonder whether many of the well-documented early outbreaks of SBS may have reflected mass hysteria rather than any specific environmental cause. In fact, some have even questioned whether it is the workers or the buildings that are 'sick' (implying that there might be a psychosomatic component to the condition). Nevertheless, whatever the cause, anything that can be done to help reduce the level of indoor air pollution is likely to reduce the symptoms of SBS and, by so doing, increase workplace productivity (a 6 per cent increase in typing speed being reported in one study).[11]

However, beyond these major negative health outcomes associated with light hunger, poor perceived air quality and too much background noise (a topic that we'll come on to later), what else can be done to sensehack the workplace in order to help maintain our levels of alertness, reduce our levels of stress and promote creative thinking? Given the findings noted earlier in the book, there are no prizes for guessing that much research has focused on the benefits of bringing nature into the workplace. However, before we get on to that, I would like to take a look at the various basic sensory aspects of the environment that affect our performance and well-being while at work.[12] If we do not change our ways, the danger is that we will end up looking like Emma, the product of a recent report by office design specialists Fellowes.[13]

Emma, the office worker of the future. Red eyes, hunched back, headaches and a host of other health problems. This may be what many of us will look like unless we start to change our ways at work, according to the latest Future of Work report. After interviewing more than 3,000 workers from France, Germany and the UK, it was suggested that 90 per cent of office workers will suffer from such issues and have difficulty doing their job if changes to work environments are not made. Already, 50 per cent of those interviewed suffered from sore eyes, 49 per cent from sore backs and 48 per cent from headaches.

## Is the air conditioning sexist?

You can't help but notice that the genders differ as far as the right position for the thermostat is concerned. While office environments that are too warm can lead to increased tiredness, heat is not something that many women complain about. They are much more likely to have the opposite problem, namely working in offices where the air conditioning is set too low, meaning that they need to wrap up in order to feel warm. The difference in thermal comfort between the sexes is not small either. According to one study, the most striking difference is between European and North American men and Japanese women, who prefer an ambient temperature that is an average of 3.1°C (5.6°F) higher than the 22.1°C (71.8°F) preferred by the Western men. The science behind the difference is that men typically have more heat-producing muscle mass than women, and so their metabolic rates are much higher (up to 30 per cent faster, in fact). Unfortunately for female office workers, the building guidelines that were established some decades ago were based on what would make an eleven-stone forty-year-old man maximally comfortable, thermally speaking. These days, many women are no longer happy to shiver in silence. Indeed, there have been increasingly strident complaints that air conditioning is sexist. Come to think of it, it might also be considered ageist too, given that our metabolic rate tends to decline over time, meaning that older workers probably also prefer a higher office temperature.

One obvious solution would be simply to increase the mean temperature on the thermostat, though perhaps there is no happy medium as far as the temperature in the office is concerned. The battle for the thermostat is about more than just thermal comfort, though, as the ambient temperature also affects our workplace performance. In one study of more than 500 men and women, the latter were shown to perform better at mathematical and verbal tasks when the ambient temperature was higher, across a broad range of temperatures from 16 to 31°C (61–88°F), while the reverse was true for men. However, given that the benefits for women's performance of turning the thermostat up were larger than the decrease in men's performance (a 1–2 per cent increase in performance in maths and verbal tasks per degree elevation in temperature in women vs a smaller decrement of around 0.6 per cent per degree centigrade in men), the authors of the study went on to suggest that performance in mixed-sex offices would improve overall if the temperature were to be turned up (assuming, that is, a roughly even gender balance). At the same time, however, growing environmental concerns about the energy efficiency of many office buildings means that it is becoming increasingly difficult to justify spending more on heating, in spite of the evidence.[14]

Some innovative designers have considered whether 'warm' paint and/or lighting colours might reduce the heating costs in winter while maintaining thermal comfort. Indeed, there is some limited evidence that people are content with a slightly lower ambient temperature when bathed in a warm yellow light rather than cool blue. That said, it remains an open question as to quite how much of a difference such a visually induced warming effect (which equates to something like a 0.4°C (0.7°F) increase) would make in a real-world setting.[15] Bear in mind, though, that the use of warm colours may have effects that go beyond just our thermal warmth, with brightly coloured paint and lighting also influencing our mood and emotion (as we saw in the Home chapter).[16]

Given that the battle for the thermostat is unlikely to go away any time soon, one futuristic solution may be temperature-controlled seats, as already found in many high-end cars. Researchers have already demonstrated that infrared monitoring devices can be used to measure the skin temperature of the occupants of an office and individually adjust the heating/cooling accordingly, and this personalized approach to thermal comfort reduced heating/cooling costs by a not insignificant 20–40 per cent. Such cost-saving interventions are obviously going to become all the more important as concerns about making our office buildings as energy efficient as possible inevitably increase.[17]

## Who hasn't felt tired at work?

When was the last time you felt sleepy at work? In a representative national survey of almost 30,000 workers in the US, interviewed in the opening years of the century, almost 40 per cent answered in the affirmative to the question 'Did you have low levels of energy, poor sleep or a feeling of fatigue in the past two weeks?' Our alertness varies predictably over the course of the day given our circadian rhythms, with many of us tending to feel sleepy late in the morning and in the afternoon, especially if we have just had a heavy lunch. According to the psychologists, we perform best at moderate levels of alertness. The key question here, then, is whether environmental cues can be used to help modulate, or manage, our levels of alertness.[18] Exposure to bright lighting and/or background music have both been shown to help. For instance, brighter artificial white (i.e. polychromatic) light typically leads to improved subjective alertness regardless of the time of day. One report in the aptly named journal *Sleep* in 2006 stated that a dose of bright afternoon light (something you might equally well get from a special lamp as from the great outdoors, assuming that the sun is shining) helps tackle sleepiness after lunch.[19] Meanwhile, playing background music has been shown to help reduce boredom and improve the productivity of factory workers and typists by as much as 10–20 per cent. That said, there are always challenges around finding something that everyone likes to listen to. This is why some recommend the use of personalized music via headphones wherever possible.[20]

Similarly, exposure to an alerting ambient scent such as peppermint or citrus may provide an effective means of pepping us up, while relaxing scents such as lavender can help to calm us down should we be feeling stressed. Talking of which, one of the other top tips after a stressful meeting is to change the scent at your desk. This simple sense hack will likely help you to reset mentally. However, be aware that the one thing that a pleasant ambient scent such as citrus simply cannot do is to make your cluttered desk look any less untidy (much though my wife would wish that it were otherwise).[21]

What many of us really need is a workplace environment that helps us not only to maintain our alertness after lunch but also to relax at the end of the day as we get ready to go home. Such constantly changing demands upon our environmental stimulation partly explain why fixed attributes of the environment are of only limited use in helping us to manage our well-being and productivity at work. So, while a particular paint colour may help to keep you alert, that might not be what you want just before you finish up at the end of the day. Intelligent lighting solutions provide a more flexible way to vary the pattern of ambient stimulation over the course of the day. In fact, some of the most intriguing work in this area has been on the blue light of dawn that we came across in an earlier chapter. Exposure to such short wavelength (460nm) blue light, even for relatively short periods of time, boosts alertness and cognitive performance across a host of different tasks.[22] Potentially perfect, then, if we are feeling sleepy.

I have spent much of the last fifteen years working with the paint and fragrance industries trying to assess the impact of the environment on people and helping design multisensory strategies to enhance their workplace performance. 'Are there specific paint colours that would make workers more productive?' was one of the questions that I investigated for Dulux Paints almost twenty years ago. There have been many studies into the effects of colour on various aspects of our mental performance.[23] In Oxford, while we tried long and hard to prove that certain paint colours would make people significantly more productive in the workplace, we found that varying the hue and brightness of the ambient lighting or screen was nearly always a more effective sense hack. Brighter lighting unsurprisingly leading to a more arousing environment.

At the same time, it has proven challenging to replicate some of the more headline-grabbing effects of staring at coloured computer screens (such as Mehta and Zhu's claim that staring at a red screen improves spell-checking while staring at a blue screen fosters creative problem solving that appeared in the pages of *Science* in 2009).[24]

## Sensehacking creativity

Talking of creativity, I must admit to despairing of many of the business innovation workshops that I am invited to. I confess that I have spent far too much of my consulting time cooped up in windowless basements in more or less glamorous hotels, sometimes for days on end. White walls, angular surfaces, no windows, no natural lighting. Certainly no sign of nature but for the occasional sad-looking pot plant abandoned in a corner. Whoever thought that such environments facilitated innovative thinking? Too many of those in charge of organizing such meetings would seem to be operating under the mistaken belief that if you get the right people together for long enough then the environment itself does not much matter. They are very much mistaken.

The physical characteristics of the places in which we work influence the way we think more than we realize. And while many of the effects, when considered individually, might be small, every little helps. Taken together, they may ultimately have a big impact on how we perform. In the Home chapter, for example, we already came across the suggestion to sit people at a round table if you want them to come to an agreement, and to make sure that the room has a high ceiling if you want to encourage lofty thoughts and have people connect ideas.

I like to contrast business innovation meetings with the creative spaces that one often finds in trendy ad agencies, and latterly Silicon Valley tech companies. One of my close friends at university went on to work in the iconic building in LA that is modelled on a pair of binoculars. When I once went to visit him, more than a quarter of a century ago as a poorly paid young academic, I was very envious. I remember being struck by the dedicated space that had been set aside in the office specifically for ideation and creative thinking. Billowing floor-to-ceiling white curtains gently rippling in the warm Californian breeze, soft white cushions so large that anyone sitting in one would be immediately swallowed up, and silence. It was all too easy to imagine how the creative process would have been enhanced with such a change of circumstances (and the evidence would appear to corroborate this).[25] That said, those in charge might not have been altogether right about the silence, for Rui Mehta and colleagues have reported that having a little background noise (described as a blend of multitalker cafeteria noise, roadside traffic and distant construction noise) can sometimes facilitate creative cognition. Across five experiments, these researchers found that noise presented at 70dB (think of the noise made by a shower or dishwasher for comparison) facilitated performance relative to when the same soundtrack was presented as 50dB or 85dB.[26]

Sometimes these things can go too far. I, for one, have yet to be convinced that the elevated gondola situated over a fake ski slope that employees working at Google's Zurich office can apparently use for informal meetings is a good idea. Note here how the rise of shared office spaces, as offered by companies such as WeWork, also tend to incorporate trendy features rarely found in standard offices.

One of the essential features of most meetings are drinks, with coffee, tea and other caffeinated beverages seemingly ubiquitous. Indeed, survey results suggest that the majority of workers expect drinks to be provided in meetings. But why?

Can a hot drink really improve our reasoning abilities? While many of us rely on stimulants such as coffee or cola to help perk us up during the working day, collaboration in group settings is enhanced by a hot and/or caffeinated drink. What is more, for those who are trying to cut back on their consumption, the good news is that our mental performance can even be improved by smelling the aroma of coffee alone, at least if we believe that the smell will alert us.[27] Sensehacking the chemical senses is much more important than you might think.

## Open-plan offices

One of the biggest challenges facing many workers is the move from private cellular offices, occupied by three people at most, to the increasingly common open-plan office format. More than 70 per cent of workers in the US now find themselves thus situated.[28] It is often claimed by those who make the decisions that the format both reduces costs and enhances co-worker interaction. However, the evidence mostly shows the opposite to be true. Time and again, the move to an open-plan office has been linked with increased levels of stress, *decreased* interpersonal interaction and a lowered feeling of subjective well-being.[29] Fatigue, headaches and stress-related illness all typically increase as a result of moving to an open-plan office. As one of the systematic reviews of the health and performance of workers put it, 'there is strong evidence that working in open workplaces reduces job satisfaction'. No wonder the newspapers so often carry articles providing sense hacks for dealing with the problem of office distraction.[30]

I recently experienced the downside myself when my own department was closed suddenly following the discovery of high asbestos levels. Most of the faculty members were moved temporarily from our sole-occupancy offices in the old building to a new open-plan location. The negative consequences were clear to see, with student researchers working far more off-site, typically citing noise distraction as a key problem. Based on my own brief experience, I certainly have no trouble believing the figures suggesting that people lose eighty-six minutes a day due to disturbance when trying to work in an open-plan office. Along with many of my colleagues, I soon started working from home too. In fact, this is where I have ended up writing much of this book! It really was the only way to get a bit of peace and quiet, not to mention privacy.

However, not everyone is fortunate enough to have that option. Sadly, the open-plan format is becoming increasingly common across the university sector, as academia slowly follows the lead of big business in developing so-called 'academic hubs'.[31] Ideally, of course, we would simply eliminate open-plan offices altogether and improve staff well-being and productivity in the process. However, something tells me that this is just not going to happen anytime soon. The short-term cost savings and increased flexibility of the open-plan format are just too hard for the bean counters to resist.

## Sensehacking the open-plan workspace

Should you be unlucky enough to find yourself in an open-plan office, the first thing to do is to try to choose a desk that is as close to a window as possible. It really can help maintain your levels of satisfaction. You should also go for a high partition separating you from your office mates, if available, so as to reduce any peripheral visual distraction.[32] One of the biggest problems with open-plan offices, though, is noise, especially from other people's conversations. Absolute silence is almost as bad, as that can all too easily seem oppressively library-like. And with 25–30 per cent of those working in open-plan offices dissatisfied with the level of noise, what are urgently required are sense hacks that can help to alleviate the problem of sonic distraction. In his book *The organized mind*, Daniel Levitin starts with a pair of earplugs, which can cut background noise by as much as 30dB. He also recommends noise-cancelling headphones to mute the background chatter still further. Less subtly, he suggests advising your colleagues not to interrupt you and telling anyone who talks too loudly to shut up! You probably don't need me to tell you what such an approach will do to your popularity ratings though.[33]

One of the other solutions that is already being used in this area involves pumping brown noise into the workspace.* The trick here is to present it at a level that is loud enough to mask speech while at the same time being quiet enough that people do not have to raise their voices to be heard. That said, spending all day listening to something that sounds rather like a mistuned radio or possibly ventilation noise is, I imagine, not to everyone's taste.

Another innovative solution that has been trialled involves the use of nature sounds. Given what we saw in an earlier chapter, there would seem to be grounds for believing that this ought to work. Relevant in this regard, a 2017 Finnish study evaluated the consequences of playing one of four different natural water soundscapes compared to regular brown noise for those working in an open-plan office. The nature soundscapes were designed to evoke a waterfall, a gentle river, a babbling river and a river with

* Brown noise is filtered pseudo-random noise that one hears with a mistuned radio or TV, say, but with the frequencies chosen to match those of human speech rather than being evenly distributed across the entire frequency spectrum, as in the case of white noise.

occasional birdsong. Each auditory masking condition was presented for at least three weeks to the seventy-seven employees, with all of the sounds being played at roughly the same low level of 44dB (i.e. not much louder than the sound of a babbling brook). Contrary to what might have been expected, the nature sounds did not result in greater subjective acoustic satisfaction, or reduced distraction, as compared to the original brown noise already in use in the office. In fact, across a range of subjective satisfaction and performance measures, the brown noise came out top as far as acoustic masking was concerned.[34] Hence, to the extent that such results are representative, they really do suggest that brown noise, unappealing though it sounds, may actually be the best sense hack in terms of reducing the stress associated with the auditory distraction present in most open-plan offices.

However, this result does leave us with a niggling doubt/question, or at least it does me. Why didn't the natural water sounds work in this situation, given the positive response that they have been shown to elicit elsewhere? One problem may simply be that playing the sounds of flowing water in an office environment is in some sense incongruent. It may make those who hear it think that there is a leak, or a problem with the toilets, rather than associating the sound with nature. (Either that, or they feel like they need to pee all the time!) By contrast, when such natural water sounds are played outside, as in those studies where they were used to mask traffic noise in park settings, for example, they often tend to work much better (as we saw in the Garden chapter). This is presumably because they are more congruent with the natural outdoor surroundings, or at least they may be treated as such by those who hear them.

The mental image a sound creates is as important as the physical properties of the sound waves themselves in explaining how background sound/noise affects us. To illustrate the point, let me tell you about a 2016 study in which Swedish researchers played the same ambiguous pink noise with interspersed white noise to three groups of people.* To one group, the experimenters said nothing. A second group was told that they could hear industrial machinery noise, while a third group was told that they were listening to nature sounds, based on a waterfall. Intriguingly, subjective restoration was significantly higher amongst those who thought that they were listening to the nature sounds than in those who were told that they were listening to industrial noise. And, as one might have expected, the results of the control group, who had been told nothing about the source of the sound, fell somewhere in between.[35]

## Bringing nature to the workplace

Those office workers who are able to get out into nature, even for just a few minutes at lunchtime, say, will perform better afterwards. Exposure to nature not only helps reduce our stress levels but also enhances our creative problem solving when we return to our desks. It is, however, important to recognize that not everyone has either the time or the opportunity to access nature during the working day. For those people, the next best thing is to site your desk near a window, preferably one with a view of nature, not to mention lots of natural light. Both of these factors exert a profoundly positive effect on both our subjective well-being and our recovery from stress.[36]

That too, though, is not an option for everyone, so what else can be done to bring an element of nature into the workplace? Traditionally, the answer was a few potted plants or perhaps some nature posters. Anger and stress levels amongst male

---

* Pink noise has the same intensity at all frequencies, whereas white noise tends to sound harsher, as the higher frequency sounds are louder.

employees do tend to be lower in offices with landscape nature paintings and/or abstract art posters on the walls.³⁷

As to the long-debated efficacy of shrubbery (i.e. potted plants) in the workplace, a 2014 study provided some of the most convincing evidence to date supporting the beneficial effects of a green, as compared to a lean, office. In a trio of field studies (excuse the pun) conducted in large commercial open-plan offices in the Netherlands and the UK, workers' subjective levels of satisfaction with their environment, as well as more objective measures of their productivity, were compared in green and lean versions of the same office. The results highlighted a clear benefit for green over lean on both subjective and objective measures of performance.³⁸ From the employer's perspective, workers were almost 25 per cent faster in the green office. Meanwhile, the employees themselves reported feeling that the air quality was better, and that their self-reported ability to concentrate was also higher. So, a win–win situation.

## *'What's wrong with plastic trees?'*

So asked a 1973 article appearing in *Science*. The author rather apologetically went on to answer his own question as follows: 'My guess is that there is very little wrong with them. Much more can be done with plastic trees and the like [than real ones] to give most people the feeling that they are experiencing nature.'³⁹ But does plastic shrubbery really provide all the same benefits as exposure to living plants? While the former might well convey some kind of psychological benefit,* what plastic plants are unquestionably less good at doing is helping to purify the air. After all, indoor plants and/or the root-zone microorganism

---

* That said, I have yet to see a well-controlled experiment on this issue.

microcosm (basically all the life that is lurking in the soil) help remove a number of the volatile organic compounds that are the suspected cause of many of the indoor air-quality-associated health problems and symptoms of sick building syndrome.

Indoor plants and/or the associated root microcosm can also help refresh the air by absorbing carbon dioxide from the atmosphere. In one 2007 study, for instance, indoor plants reduced carbon dioxide levels by as much as 25 per cent in naturally ventilated buildings, and by 10 per cent in air-conditioned offices. There is some doubt, though, as to whether it is the leaves themselves or the bacteria that either live on them or are found in the soil that are actually doing the work here. And should you be wondering which varieties to go for, the suggestion is that small, green, lightly scented plants tend to work best in terms of enhancing our health and well-being. Watch out for red-flowered plants in the office, though, for while they may well be visually attractive, they can apparently become fatiguing after a while.[40]

I well remember the look on my students' faces when I turned up at the lab one day with a car full of shrubbery. They were clearly of the opinion that I was engaged in some sort of psyops, introducing the pot plants not for aesthetic reasons but as a calculated means of increasing the number of publications that I could extract from them. Maybe they were right to be worried! After all, the potential performance-enhancing benefit is not to be sniffed at. According to the World Green Building Council, productivity increases of between 8 and 11 per cent can be gained simply as a result of increasing ventilation and reducing pollutants in the workplace.[41] Such impressive results perhaps help to explain why, when Amazon opened its new flagship offices in downtown Seattle early in 2018, they looked more like a greenhouse than a regular office building. The Spheres' three glass domes house some 40,000 plants of 400 species, as well as

Amazon's flagship offices in downtown Seattle.

Amazon HQ!⁴² If it is possible to overdose on the nature effect, then this surely is how.

## All the beauty of nature on your desktop?

Many office workers spend more time staring at a computer screen than at anything else. The question therefore arises as to whether what is seen there might also trigger the nature effect. (Think of the often stunningly beautiful landscape wallpapers and screensavers found on Windows.) Whenever I come back to my computer after a break these days, I am greeted by a seemingly endless array of gorgeous images of nature. Does it, I wonder, benefit me to stare at such images for a few minutes every now and then? Might it even help to restore my attentional resources and/or perhaps enhance my recovery after a tense and stressful work meeting or, more likely in my case, after receiving the email notifying me that my latest grant application/academic paper has just been unceremoniously rejected? I suspect that it does, at least if I stare at such images for long enough.* The key question here, of course, being how long that might be.

The closest one gets to assessing this comes from research where people have been given cognitive/attentional tasks to perform both prior to and after viewing a series of putatively restorative nature images on a computer monitor. In order to control for the effects of practice the second time round, any performance improvements in the nature condition are normally compared with those observed in another group of participants who get to stare at urban scenes, or else perhaps something neutral like geometric patterns. The results of several such studies have now confirmed that viewing nature images on a computer screen† can indeed help to restore people's attentional abilities, as evidenced by the performance improvements seen in various standardized tests.[43]

As yet, the minimum amount of time that one needs to look at such images in order to demonstrate a significant benefit in terms of attentional restoration hasn't been precisely worked out. However, in one study a measurable improvement in performance was documented after no more than forty seconds viewing a picture of a green flowering roof garden rather than a concrete-covered roof.[44] That said, quite how much benefit one gets from viewing nature images may well depend both on how long you look at them for and also on the size of your screen. That, at least, was the conclusion from another study

---

* Remember here the two popular accounts of what nature might be doing to us, one originally proposed by Ulrich in terms of recovery from stress, and the other, attention restoration theory, as championed by Kaplan and his colleagues. As we saw in the Garden chapter, these two theories need not be treated as mutually exclusive.

† For 6¼ minutes in one study and 10 minutes in the other, with each image being shown for somewhere between 7 and 15 seconds.

showing that the bigger the monitor, the more immersive the viewing experience, and the more restorative in terms of people's recovery from a mild stressor. In this case, the subjects were stressed by having to perform an arduous arithmetic test for sixteen minutes while listening to industrial background noise.[45]

While the idea of staring at a nature-scenes screensaver on a supersize monitor is all well and good, there is, of course, work to be done, and most of the time your computer screen is going to be occupied by something far less inspiring. How else, then, can the benefits of nature be brought into the workplace for those who are unable to sit by a window? One of the intriguing suggestions that has been put forward in terms of hacking the senses is to place a virtual window on the wall (i.e. a screen showing a live feed of nature). Might that be as good as a real view in terms of improving our well-being in the workplace? This question has been assessed by researchers from the University of Washington in Seattle. They compared the effects on cognitive restoration after a mild stress test of a window, an HDTV feed and a blank wall. The results were clear: the participants' heart rate returned to baseline levels much more rapidly in those looking through the window. Disappointingly, the TV display showing essentially the same scene was no better than the blank wall.[46]

As we know, nature is nothing if not multisensory, and one of the most obvious limitations with a screen view is that it usually only captures the sight of nature, not its sounds. Consistent with this view, the results of a 2013 pilot study (meaning that only a small number of participants were tested) demonstrated that people recovered from the Trier Social Stress Test (see the Home chapter) much more rapidly if exposed to an audio-visual virtual reality rendition of nature).[47] Once again, it turned out that displaying a nature view virtually (in the form of a silent forest) was no better than staring at a blank wall.* One conclusion to draw from this study, and the one preferred by the authors, is that experiencing nature with more of your senses is better.†　It also leaves open the unresolved question of when, exactly, just staring at a digital rendition of nature helps – remember the screensaver study we came across a moment ago.

Looking further into the future, I think that it would also be interesting to incorporate the smell and feel of nature into the workplace too. After all, scents such as citrus and peppermint have been shown to improve not only people's mood but also their performance across a range of different tasks.[48] Would the smell of the forest floor help the participants in these laboratory studies to recover more effectively? I suspect that it would. Or maybe a sprinkle of geosmin, the key volatile found in petrichor, familiar as the smell of the dry earth after it rains.

And why not drape a textured throw over your office chair? At the very least, a throw is likely to absorb some of that background noise. Something that I like to do is to place a natural object on my desk – a stone, a pine cone, a chestnut or a piece of tree bark; something with a natural feel to contrast with all the artificially smooth surfaces that fill my office environment. I doubt that touching nature will ever turn out to be as good for us in terms of enhancing our subjective well-being as seeing or hearing it, but it is a start. As we have seen throughout this book,

---

* Intriguingly, a few of the participants even said that they found the silent forest condition a little threatening. They had the feeling that something bad was going to happen.

† A critic might, though, point to the fact that the researchers failed to assess performance in the nature sound only condition. And if you were wondering why the Scandinavians have been so active in the area of multisensory office design, blame all those long dark cold winter nights that tend to exert an especially negative impact on the well-being of workers over there.

the biggest benefits to our health and well-being tend to result from combining the senses in a congruent manner.[49]

## What is the connection between creativity and commensality?

It is probably no coincidence that a number of Silicon Valley's most successful tech companies – think Google, Pixar, Apple, Yahoo and Dropbox, to name but a few – have at least one thing in common: they all offer subsidized, or in some cases free, food to their employees. What is more, this is often eaten at long communal tables (a bit like an Oxbridge college dining hall, come to think of it). There is obviously a substantial cost associated with such largesse, but one that is carefully assessed by those in charge. At least that was the line from Michael Bakker, Head of Food at Google, the last time I spoke to him at a hospitality-industry conference. Such an arrangement, note, encourages chance encounters by having people who do not necessarily know each other meet, be it at the dining table, or perhaps at the coffee stand.* Such changes to the design of innovative workspaces in recent years have coincided with a growing realization of the value of knowledge creation.[50]

One commentator writing in *Forbes* magazine suggested that the strategic reason behind all that free food at Google 'isn't just to trick employees into staying on campus. Its purpose is actually to inspire innovative thinking.' The aim, in other words, is to make people interact![51] Problems can soon arise, though, if one's food offering becomes too popular. After all, who amongst us enjoys queuing for ages in the staff canteen. Intriguingly,

---

* Some now talk of the 'coffice', in light of the increasing tendency for the coffee shop to become the sociable meeting place for today's millennial workers.

Shimazu, the designers of many Japanese office buildings, have come up with an innovative solution to this particular problem. They pump out food smells through the ventilation system on different floors of the office building at different times in order to manage the mealtime rush in their small canteens. Given how often our appetite is triggered by food aromas, this is likely to be an effective strategy.

Those who eat the same food are more likely to cooperate in a trust game or labour negotiation scenario than those who are given something different to chew on. As such, the provision of food can also be used strategically to facilitate business negotiations. In research by Lakshmi Balachandra of Babson College (US), 132 MBA students were given the task of negotiating a theoretical complex joint venture agreement between two companies. They came to settlements that were 11–12 per cent, or $6.7 million, higher for the two parties concerned when food was involved than when it was absent from the negotiating table. It is not just creative types who benefit from eating together, at least not if the research by Brian Wansink, formerly of Cornell University, and his colleagues is to be believed; they found that commensality (i.e. eating together) at mealtimes improved the performance of firefighters.[52]

Some enlightened politicians are now starting to take food far more seriously too. For instance, Hillary Clinton was certainly interested when, as Secretary of State, she ushered in a whole new approach to food provision, as a part of what she termed 'smart diplomacy'. Indeed, as Natalie Jones, a deputy chief of protocol in the US government put it, food is crucial 'because tough negotiations take place at the dining table'.[53] Meals with visiting heads of state and other dignitaries were apparently used by Clinton as occasions to showcase North American cuisine and local produce while at the same time highlighting sensitivity to foreign tastes and customs. The hope was that this would help cultivate a stronger cultural understanding.

One should never neglect the fundamental role that eating together can play in the workplace. And while there might be no such thing as a free lunch, failing to consider the role played by the chemical senses in creating the most successful working environment, especially for those working in the knowledge economy/creative industries, is a mistake that employers would do well to avoid. So, when thinking about what to do, why not take a leaf out of the book of some of the most successful businesses on the planet and universities like Oxford and Cambridge, where the food provision for staff is also heavily subsidized. They did not get where they are today without thinking carefully about every aspect of the multisensory environments that their staff work in. However, that's enough about work. For, as the old proverb (traced back as far as 1659) says, 'All work and no play makes Jack a dull boy.'

# 7. Shopping

Who hasn't had the experience of going shopping for one thing and coming home with a whole host of other purchases that you had not intended to buy? Or else, of clicking on all manner of goods online only to find yourself returning many of them after you realize that you do not really need them? Perhaps you should not feel quite so guilty though, as this might not be all your own fault. After all, one of the places where sensehacking has become much more of a science than an art is in the world of retail. For years now, companies have been using the latest findings from the emerging fields of neuromarketing and sensory marketing to tempt us.[1] Once they have lured us inside, or onto their website in the case of online shopping, they do everything in their power to make us linger for a little longer. Their hope is that we can be encouraged to buy more than we need and be nudged towards more expensive purchases than might otherwise have been the case. After all, get the multisensory atmosphere right, and/or deliver a fluent processing experience online, and it would seem that we can all be turned into hopeless, or should that be helpless, shopaholics.

I am afraid to say that there is little that can be done to stop these subtle, and not so subtle, influences on our behaviour, whether we are aware of them or not. The evidence suggests that most of us simply refuse to believe that we can be so easily swayed; we all believe that, while others might fall prey to such cheap tricks – 'half-price offer', 'buy one, get one free', 'for a limited time only', you know the kind of thing – we certainly wouldn't. Nevertheless, I think that we should all be a lot more concerned about the hacking of our senses in the marketplace than we currently are. And I should know, having spent much of the last quarter of a century working with companies and advertising agencies, both large and small, to make you buy more of everything from deodorant to detergent, and from coffee to clothing.[2] I am certainly not the first to raise concerns about the hidden power of marketing. In 1957, Vance Packard's *The hidden persuaders* rapidly became a classic piece of journalistic scaremongering along just these lines.

The marketers' understanding of how to titillate our senses while shopping has certainly come a very long way since the early days of motivational research by the likes of Louis Cheskin and Ernest Dichter, described in Packard's book. These early practitioners were amongst the first to realize that they could bias consumers' perception and behaviour through the use of abstract colours and shapes on logos, labels and product packaging.[3] For those of you who haven't heard of him, Cheskin is popularly credited with introducing the red circle to the middle of the 7UP logo. You have probably never thought about what it

The 7UP logo. But what exactly is the red circle in the middle saying?

is doing there, have you?* Cheskin also convinced McDonald's that it was a good idea to hold on to its Golden Arches.

Nowadays, consumer neuroscience, the preferred name for neuromarketing amongst many academics, allows researchers to peer directly into the shoppers' brain in search of the haloed 'buy button'. This without the need to rely on what people say.[4] What is more, machine learning and big data analysis are also starting to provide some intriguing insights into the drivers of human behaviour that go way beyond anything that Cheskin, once hailed as Madison Avenue's marketing magician, could ever have dreamed about in the middle decades of the last century.[5]

## *Leading the customer by the nose*

Let us start, though, by considering the supermarket, as this is where much of the research has been conducted to date. In part, this is because food stores present ideal targets for multisensory marketing interventions. One of the reasons is that customers' senses are stimulated by so many of the products on display. What is more, the large number of low-cost repeat-purchase items can also provide a rich source of data. This is something that those who control the loyalty-card schemes know only too well.† (I cut all my loyalty cards up years ago.) First to mind when most of us think about sensehacking in the supermarket are the (supposedly synthetic) bread smells that are pumped out by many stores.[6] Intriguingly, however, as far as I can tell, no research has been published on the topic. It is not that the research hasn't been done,

---

* Don't worry, I'll tell you later.
† Curiously, this is not always the supermarkets themselves. I know of a number of supermarket chains who have no automatic access to their own customers' loyalty-card sales data, hard though this may be to believe.

you understand, for it most certainly has. It is just that the supermarkets have chosen not to publish their findings. Off the record, though, a number of industry sources have confirmed to me that they are sitting on data showing the dramatic effects that such ambient aromas have on sales.

The one thing to say straight away here is that the bread smells that titillate your nostrils in the store are unlikely to be artificial. This is because the delectable smell of freshly baked bread is one of the aromas that, at least until very recently, chemists struggled to imitate synthetically. Of course, just because the smell is probably 'real' doesn't mean that it hasn't been cleverly vented so as to hit you just as soon as you walk through the doors, or even when you happen to pass by on the street outside. According to a report in the *Wall Street Journal*, when searching out new locations for their outlets, chains such as Panera Bread, Cinnabon and Subway often home in on those spots located close to the bottom of stairwells in shopping centres. The idea is that this will help their distinctive scents to travel further. Not only that, these chains typically use the least powerful extractor hoods that they can get away with. And that is all before you read about the Cinnabon stores baking sheets sprinkled with nothing but powdered cinnamon and brown sugar, to make sure that there is an appetizing aroma for anyone in the vicinity.[7] So, given all the evidence, I would say that these companies really do appear to be using smell for the hard sell.*

Dutch researchers documented a 15 per cent increase in sales when they diffused a synthetic melon scent through a supermarket.[8] Not a bad return on investment, even once the cost of installing all that scent-enabling technology, not to mention the

---

* And why is it that you always seem to be greeted by the perfume displays as soon as you walk into a department store? Is this perhaps another example of olfactory marketing?

constant stream of synthetic refills, has been factored in.[9] So, next time you find yourself salivating in response to all those intoxicatingly sensual aromas of fresh produce and baked goods as you wander through the food hall of Le Bon Marché in Paris, or Dean & DeLuca in the States (before it went bust, replaced by the likes of Whole Foods and Eataly), or even Tesco in High Wycombe, you might just want to stop and ask yourself whether you are being led by your nose into making purchases that you otherwise might not have done. I am certainly convinced that the ambient smells that surround us in many public spaces exert a much more powerful influence over our shopping behaviour, not to mention our waistlines, than we realize.

This differs markedly from the situation I found a few years ago when I was invited to do some work for Thorntons, a famous chain of chocolate shops in the UK. Walk into any one of their outlets, close your eyes, and inhale. What would you expect to smell – chocolate, right? Bizarrely, in this case, the answer was absolutely nothing. You really could have been anywhere – a mobile phone shop, perhaps? All of the chocolates in the gift boxes and mixed assortments were hermetically sealed behind cellophane wrappers, a real lost opportunity as far as olfactory marketing was concerned. Contrast this only with the chocolate aroma that is present throughout what was, for a while, the world's largest sweet shop: the M&M's World store in London's Leicester Square.[10] This lack of in-store scent in Thorntons was all the more surprising when you consider that chocolate is one of the world's most desirable aromas.[11] My advice fell on deaf ears (or should that be 'blocked noses'), and I was not surprised to hear that the chain had been taken over by the Ferrero Group, makers of Ferrero Rocher chocolates, in 2015, with store and staff numbers in steady decline.

Coffee is another of the world's most-liked aromas. It is widely used in retail, and not just by those wanting to sell more of the popular beverage either – just think, for example, of the tie-up between Starbucks and both Barnes & Noble bookshops (in the US) and Uniqlo clothing.[12] According to one industry report, coffee sales at the service station can be more than tripled simply by spritzing drivers with the synthetic aroma of freshly ground coffee while they are filling up at the pumps out on the forecourt.*[13] This approach to marketing is undoubtedly one that my grandfather would have endorsed. After all, he was known to sprinkle a handful of aromatic coffee beans on to the floor behind the counter of his greengrocer's store first thing in the morning. As he walked over the beans serving the customers, their nostrils would be tantalized by the smell of quite literally freshly ground coffee. This is one of the examples mentioned in my last book, *Gastrophysics: The new science of eating*. My grandfather would therefore intuitively seem to have devised an effective sense hack designed to boost sales long before the contemporary interest in 'scent-sory' marketing emerged.

A very contemporary twist to this kind of approach was provided by a prize-winning 'Flavor Radio' campaign executed by Dunkin' Donuts in Seoul, South Korea. Intelligent scent dispensers were installed on a number of the city's buses that would recognize when the Dunkin' Donuts jingle was playing on the in-vehicle radio, and respond by releasing coffee aroma. The idea was that after stepping off the bus, passengers would soon stumble across one of the chain's stores and make a purchase. The evidence suggests that this multisensory marketing strategy really did work, with a 16 per cent spike in visitors to Dunkin' Donuts branches situated close to a bus stop, as well as a 29 per cent increase in sales of coffee.[14] No wonder that this innovative campaign picked up one of the much-coveted prizes at the

---

* Such industry-sponsored findings should probably be taken with a hefty grain of salt, given that independent olfactory research rarely shows such large effects.

prestigious Cannes Lions festival – one of the top annual awards ceremonies for the creative industries. That said, it is an entirely separate question as to just how cost-effective (and, arguably, ethical) the campaign was.

### Smelling colour

When we smell something distinctive such as a strawberry, we tend to look preferentially at those objects that we associate with their source, and pick out objects much faster when in the presence of their distinctive scent or a sound associated with them.[15] You can see, then, how an ambient scent or background music can do much more than merely influence our mood. Sensory marketing can also be used to direct our visual attention to a specific product or brand, possibly explaining at least part of the success of Dunkin' Donuts 'Flavor Radio'.

Not everyone agrees with such practices. This was certainly the case when the 'Got milk' campaign in California started scenting bus shelters in the state with the aroma of cookies in order to give their advertising a multisensory boost. The campaign had to be pulled within days of its launch over concerns that it was grossly insensitive to the many hungry homeless people in the state who used the shelters to sleep in.[16]

Another olfactory marketing campaign that was pulled after only a day involved the Disaronno liqueur brand. Some bright spark apparently thought that it would be a good idea to flood the London Underground with the drink's distinctive aroma – the liqueur itself has an amaretto taste. Had this two-week 'scent-wafting' campaign gone ahead as planned, an almond scent would have been pumped into the ventilation system, thus appealing to the noses of all those who found themselves on the Tube. Rather unfortunately, this coincided with the release of an article in the country's most widely read newspaper, the *Daily Mail*, describing the tell-tale signs of terrorist activity. The piece warned commuters, especially those on the Underground, to be extremely cautious should they smell almonds, since cyanide is made from almonds, just like the drink! How unlucky can you be?[17] Almost as unlucky as one famous brand of Mexican beer when the coronavirus hit the world stage, you might say.

Given these public relations disasters, you will understand my relief when no such problems occurred when one of the London-based marketing agencies that I worked with a few years ago decided to send a taxi around the capital's streets pumping out the smell of McCain's Ready Baked Jackets, basically frozen microwavable oven-baked potatoes. There were even a few 3D video signs that would surprise any unsuspecting commuters waiting at bus stops by emitting that 'ever-so-moreish' baked potato aroma.* Or, as one commentator described it, 'Each billboard includes a fibreglass potato sculpture and a mysterious button: Push it, and the tuber discharges the aroma of "slow oven-baked jacket potatoes".'[18]

It does not stop there, though, for one group of Italian psychologists reported that the size of the object we associate with a specific odour can influence our reaching behaviour too. They found that when we smell something small – think of a clove of garlic or a pistachio nut – then our motor system is automatically primed to pick up a small object. By contrast, when we smell something larger, like an orange, our hands will find it just that little bit easier to grasp a larger item.[19] That said, I have yet to see or, better said, smell any cunning marketers incorporating the latter findings into their prize-winning campaigns just

---

\* So, once again, an aromatic bus shelter! Though, contrasting with the Californian example mentioned a moment ago, the scent-release in this case required active engagement, meaning that it would not torment anyone should they choose to lie down there for the night.

yet. Perhaps releasing the smell of peanuts in a jewellery store might be as good a place to start as any.*

## Moving to the beat

It is not just food aromas in supermarkets that we should be worried about. While few of us realize it, our behaviour is often entrained to the musical beat. In a now-classic study, Ronald E. Milliman, a marketing professor from Loyola University, New Orleans, monitored the flow of people in a supermarket in an unnamed south-western US city and analysed the till receipts. In one of the biggest studies of its kind, extending over a period of nine weeks, shoppers were shown to spend 38 per cent more when slow as opposed to fast music was played (60 vs 108 bpm). Such findings have been taken on board by many chains subsequently, though you will not be too surprised to hear that few of them are willing to talk about what they are up to. One of the few examples that did make it into the public domain concerned Chipotle, the Mexican-grill chain.

In subsequent research Milliman went on to show that people also eat, drink and, most importantly, spend more when slow as opposed to fast music was played in a restaurant (the idea being that slow music made diners linger for longer). According to a report that appeared in *Businessweek* magazine, Chipotle carefully controls the tempo of the music that is beamed out to all 1,500 of its stores: they deliberately play faster music at busy times of day, to try to speed up their customers, so hopefully shortening the queues, while at the same time freeing up tables. By contrast, slower tracks are played at quieter times, thus encouraging customers to remain so that their stores don't look

---

* Just joking!

too empty. According to Chris Golub, the chain's in-house DJ, 'The lunch and dinner rush have songs with higher BPMs because they need to keep the customers moving.' Golub apparently sits in one of the chain's many branches in NYC, watching how the customers respond to the music that he is planning to add to the playlist. If he sees them bobbing their heads, or tapping their feet to the beat, then he knows that he is on to a winner, and the track gets added to the playlist.[20]

If this sounds a little bit like elevator music, or Muzak, it is. Muzak was the name given to a distinctive style of easy-listening background music, very often instrumental, that would play continuously in public spaces such as shops, airports, hotels and even brothels to help customers relax.*[21]

The tempo of the music should not be considered in isolation, though, as my former post-doc Klemens Knoeferle and his colleagues demonstrated in 2012, when they highlighted the interacting influence of musical tempo (fast: in excess of 135 beats per minute vs slow: lower than 95 beats per minute) and musical mode (major or minor) on shopper behaviour. They found that Milliman's claim regarding slow-tempo music increasing sales was only true for music in the minor mode. The effect on sales of music in the major mode wasn't altered by tempo.[22]

However, my absolute favourite study, at least as far as musical manipulation in the marketplace is concerned, was first published by Adrian North and his colleagues in 1997.[23] They conducted a study in an English supermarket stocking four types each of both French and German wines, matched for cost and dryness/sweetness. For two weeks, French accordion music or German oompah-oompah music was played over

---

* Mood Media, the company that currently provides the music streaming service for Chipotle, was formerly known as the Muzak Corporation.

the loudspeakers, the ethnicity alternating daily. They found that 83 per cent of wine buyers bought French wine when the accordion music was playing, while 65 per cent of the bottles flying off the shelves were German when the Bierkeller music was played. More remarkable still, less than 14 per cent of those shoppers said they recognized the impact that the music had had on them. What these results, and many others like them, imply is that our choices can be radically manipulated by simple changes to the music playing in the background without most of us having any idea what is going on.

There are a few things about the results of this rightly famous study that are worth considering. One thing to point out straight away is that the sample size was very small, with sales data from just eighty-two shoppers, of whom only forty-four agreed to be interviewed. The percentages mentioned above make these results seem rather more impressive than the actual numbers warrant. Today, with the replication crisis★ convulsing the psychological sciences, it would certainly be nice to see a larger-scale replication of North et al.'s seminal research. This would hopefully also help to shed light on the question of whether shoppers are still as easily manipulated today as they apparently were during the 1990s.

The other thing to consider here is whether wine might be somehow a special case. After all, the wine aisle is commonly thought to be one of the most visually complex, not to mention constantly changing, in the supermarket (at least amongst branded products), thus helping to explain the emergence of so-called 'critter brands' – those wines that display a recognizable critter on their label: a giraffe, an emu, a toad, or what have you. While the critters may not have anything to do with the

---

★ This is the name given to the many recent failures to replicate some of the sexier findings in the field of psychology and neuroscience.

wine itself, the idea is that their presence on the label helps the confused shopper to remember, and hence to find once again, the bottle that they enjoyed so much last time. At least they can avoid asking for a tongue-twisting bottle of Eitelsbacher Karthäuserhofberg Riesling Kabinett, say, or a Piesporter Goldtröpfchen. And how would you pronounce the Hungarian varietal Cserszegi Fűszeres should you find yourself in the wine store? Try asking for a bottle of that without making a fool of yourself! And no, before you ask, I did not make any of those up.*

Unless we have tried it before, we have no way of knowing exactly what any wine is going to taste like until we open it, by which time it is too late to change our minds. In other aisles in the supermarket it is much easier to rely on familiar brands, or else use the evidence before our senses, to check the ripeness of the produce, say. This has led some marketing experts to suggest that different rules might apply to wine when it comes to influencing shopper behaviour. On the other hand, it has been demonstrated that our food choices can also be influenced by the ethnicity of the background music. In 2017, Debra Zellner and her colleagues from Montclair State University, New Jersey, found that playing some flamenco in a North American university canteen led to increased sales of paella while playing Italian music led to increased sales of chicken Parmesan. We also tend to spend more on wine or a meal out when classical music is playing rather than other styles such as pop. These, and many other similar findings, hint at just how much our behaviour can be affected by the sonic elements of the atmosphere. Scary stuff,

---

* According to Peter F. May's (2006) book *Marilyn Merlot and the naked grape: Odd wines from around the world*, the correct pronunciation for this crossing of Gewürztraminer and Irsai Olivér grapes is 'Chair-sheggy Foo-share-us'. So now you know.

especially when you consider that the background music is perhaps the easiest aspect of the environment to manipulate.[24]

## Subliminal seduction

At the start of my academic career in Oxford, the lab to which I was attached was approached by an international chain who wanted to know – just hypothetically, you understand – whether it would be possible to add some subliminal messaging to help direct shoppers to buy certain products. You are probably familiar with the kind of thing: 'Buy Coca-Cola' or 'Two for the price of one on own-brand laundry detergent'. The idea, in this case, was that these messages might be embedded in the background music playing over the in-store loudspeakers to help boost sales. Such a request was perhaps inspired by James Vicary's report in the 1950s that he increased sales of Coca-Cola in the cinema by briefly (i.e. subliminally) flashing up a screen suggestion 'Buy Coke' or something of the sort.*

We had to disappoint the company though. Not because subliminal sensory cues can't bias our food- and beverage-related behaviours, for they can most certainly do that. Rather, the problem was that there is only a very narrow window of opportunity in which the targeted message would have been loud enough to impact customer behaviour while, at the same time, being quiet enough for shoppers not to be aware of it consciously. Subliminal seduction, it turns out, is something that it is a little easier to demonstrate in the carefully controlled conditions of the science lab than under the noisy conditions of everyday life. In the lab, it is much easier to present the sensory

---

* It matters little what the actual message was as this turned out to be an elaborate hoax by Vicary. No such study was ever conducted.

stimuli just below each person's threshold for awareness. What is also important in terms of demonstrating subliminal priming is that the person must be in a need state (e.g. thirsty) to begin with, as shown in a Dutch study that managed subliminally to prime thirsty people to choose a Lipton Ice drink over an alternative brand.[25] And then, of course, ethical concerns are always raised when people start talking about subliminal marketing.

To my way of thinking, the red circle introduced into the centre of the 7UP logo by Louis Cheskin all those decades ago is much more interesting in this regard. Note that the red circle is not technically subliminal (i.e. it is not literally hidden), because we have all seen it innumerable times. However, I would argue that it is *functionally* subliminal, in the sense that few of us are aware of what that symbol communicates to our minds subconsciously. Note that the colour red and roundness are both associated with sweetness, like the drink itself. By showing the shape and colour that are associated with sweetness, it is possible to communicate 'sweetness', and hence prime that taste in the mind of the consumer. And, if a particular taste or flavour is primed, then we are just that bit more likely to experience it.

Take a look at the supermarket shelves and you'll soon start to see many other examples of such shape-symbolic marketing, some even predating Cheskin's arrival onto the scene. For instance, just take the stars that adorn the San Pellegrino water bottle, not to mention beer brands such as Heineken,

It's in the stars. An example of subliminal symbolic messaging.

Newcastle Brown Ale, Sapporo or Estrella. What are they doing there? The reason is that we associate carbonation and bitterness with angularity, and hence these symbols are again subliminally signalling to the consumer. It is my belief that such functionally subliminal cross-sensory marketing is a much more pervasive influence on our product expectations, product choice and subsequent product experience than any of us realize. For, regardless of the language we speak, such symbols tend to communicate their message to us at a much more universal implicit level.[26]

## *Atmospherics*

Back in 1974, Philip Kotler, the legendary North American marketer, published an influential paper on atmospherics in the *Journal of Retailing* in which he suggested that retailers should stop concentrating on just the tangible product that they were trying to sell and focus on the total product experience. He cited a number of persuasive, if largely anecdotal, observations concerning those retail establishments whose success could be put down to the total experience, or atmosphere, that the owners had managed to create. Kotler broke the store atmosphere down into its separate sensory elements, considering what could be done in terms of colour and lighting, music and scent, and even the tactile aspect of design. However, while he was ahead of his time in terms of drawing people's attention to the power of the senses in driving customer experience (and behaviour), what he, as well as the majority of the marketers who followed in his wake, singularly failed to realize is that the senses interact all the time.[27] Rarely, if ever, is the experience or atmosphere in store determined by just one of our senses. Rather, it is nearly always the combination of inputs that does the trick. The atmosphere is, in other words, multisensory, with all the opportunities and challenges that that realization entails. That being said, the majority of the research that has been conducted to date has tended to study the impact of individual cues in isolation – either just the loudness or tempo of the music, say, or the presence versus absence of ambient scent. Let's take a look at the latter.

## Smells like teen spirit

In terms of ambient scent, one of the most obvious differences between food and clothing is that there is simply no equivalent of the supermarket bread smell, except perhaps when selling leather goods. In fact, the closest one comes might well be the scent of freshly starched cotton that once was spritzed over shoppers as they passed by the relevant display stands in Thomas Pink shirt stores in New York. Anne Fontaine boutique women's clothing stores, meanwhile, add a fragrant burst of their perfume to the shopping bag while you are paying for your latest purchases – a sachet of dried flowers is normally attached to the clothes too. The sales staff at Coco Chanel's first store in Paris were encouraged to spray her perfume, Chanel No. 5, all over the boutique, from the entrance through to the dressing rooms, in order to boost sales.[28]

Meanwhile, in 2014, Portuguese brand Salsa added a matching scent via microencapsulation to their colourful jeans. According to one website, 'To help women stay sweet-scented this summer, Portuguese fashion brand Salsa has introduced its Fragrance Jeans range, which come imbued with fruity smells.'[29] The fruity fragrance of blueberry, orange, lemon, apple and strawberry apparently lingers for as many as twenty washes. Another intriguing solution to the scenting of clothes stores that I came across while working in Colombia came from clothing retailer Punto Blanco. The latter had installed a small fine-chocolates counter into a number of their stores. The aim was not necessarily to make any money from the sale of chocolates, but rather to make sure that they were engaging all of their customers' senses. Elsewhere, clothing retailers such as Uniqlo and Club Monaco have incorporated coffee shops into their stores for much the same reason, and presumably to make their customers want to hang around.

The good news for retailers is that if they get the scent right, clothing sales can be nudged in the right direction. It is important to stress, though, that there are likely to be several distinct psychological mechanisms at work here. For instance, ambient scents, just like other sensory cues, encourage approach behaviour. Then, once we have been enticed to enter the store, the pleasant scent might help to enhance our mood – the widespread belief amongst the experts being that we spend more when we are in a good mood. Then again, the fragrance might itself be distinctive enough to become a signature scent; one that is instantly recognizable and thus primes whatever associations we happen to have with a particular brand.[30] Think here only of the distinctive smell of Hollister Co. clothing stores, owned by Abercrombie & Fitch.

At the same time, a number of researchers have highlighted the importance of making sure that the scent is congruent with the product that the retailer wants to sell. And, if that wasn't enough for the poor olfactory marketer to think about, there are also cultural differences in the meaning of scents, or what we associate them with. For example, while shoppers from France and Germany consider the scent of cut grass or cucumber to be both refreshing and stimulating, those from Mexico and China apparently tend to think of these smells as natural, but not especially refreshing.[31]

That said, despite these various challenges, anyone who has

shopped for clothes in recent years can't help but have noticed the scents that are now being diffused through the ventilation system by so many stores, including Zara, Victoria's Secret and Juicy Couture. Hugo Boss, meanwhile, apparently uses an ambient scent that contains fruit and citrus notes together with a hint of cocoa.[32] Investing in scent-enabling technologies, and especially commissioning a bespoke fragrance, doesn't necessarily come cheap though. Anyone looking for a more affordable olfactory sense hack could do worse than place a bunch of fragrant flowers on the counter. According to the results of a preliminary study by Dr Alan Hirsch from Chicago, participants in the laboratory were willing to pay more than $10 more for a pair of sneakers with the scent of flowers hanging heavy in the air, while their self-reported purchase intent increased by more than 80 per cent.[33] Consistent with these observations, customers at a large metropolitan jewellery store spent more time at those counters that had been sprayed with a floral, fruity or spicy scent than at unscented counters.[34] Elsewhere, one can find a melon scent in Samsung Experience stores, while the Sony Style store apparently uses a subtle blend of vanilla and mandarin orange. Hamleys toy store in London, meanwhile, uses a scent that may remind the children's parents of nothing so much as piña colada.[35]

Lush is a popular chain of bathing and personal-care products shops scattered across Europe, North America and beyond. As best I can describe it, their signature smell is of bright clean floral notes. That said, I am never quite sure whether it is the fragrance itself that is recognizable, or merely the intensity of the olfactory onslaught assailing one's nostrils. There is, after all, nothing else quite like it on the high street. Lush cleverly sell their products without plastic or packaging in order to help their distinctive scent travel further. There is no doubt that it is easier to capitalize on olfactory marketing in some categories than others.

## *Keeping a cool head*

In terms of tactile atmospherics, perhaps the most obvious element is ambient temperature. Would it surprise you to learn that some expensive clothing stores deliberately lower the temperature in-store? How else to explain the fact that when one journalist recorded the temperature inside various clothing stores in New York, he documented an inverse correlation between temperature and price point. Luxury brands, in other words, tend to set the thermostat lower than those chains catering to the masses. As the journalist noted, 'Macy's is colder than Old Navy, but Bloomingdale's is colder than Macy's, and Bergdorf Goodman is colder than all of them . . . In other words, the higher the prices, the lower the temperature. Consider the clothing stores: Bergdorf Goodman, 68.3 degrees; Bloomingdale's, 70.8; Macy's 73.1; Club Monaco, 74.0; the Original Levi's Store, 76.8; Old Navy 80.3.' Such a thermal approach to atmospherics makes sense in light of the latest research suggesting that we tend to evaluate items more highly when the temperature is lower; or as Japanese poet Saitō Ryokuu once put it 'elegance is frigid'.[36] According to Lisa Heschong in her 1979 book *Thermal delight in architecture*, the association between low temperate and exclusiveness may have originated from the time when air conditioning was first introduced in the US, and only the boss would have the luxury of aircon in his or her office.

The shopping experience doesn't involve only sight, sound, smell and temperature, though. Physical contact amongst staff members, between staff and customers, and between the customers and the merchandise can be a powerful marketing tool in its own right.[37] I firmly believe that touch is much more important than many of us realize. And it is to this, our largest sense, that we now turn.

## 'Touch me'

Have you ever come across one of those signs in a shop saying 'Touch me'? If you did as you were told then the marketers have succeeded in nudging you towards a purchase. Even something as simple as physically picking up a product, or just imagining holding it, increases our sense of ownership, at least according to research by Joanne Peck, for many years based at the University of Wisconsin-Madison, and her colleagues.[38] Indeed, when you see the figures, it is easy to see why the marketers are so keen to get us to touch the merchandise. For instance, Asda, now part of Walmart, reported a 50 per cent increase in sales when the plastic packaging on its own-brand toilet paper was removed, so that the shoppers could feel the quality. Similarly, a large part of the success of clothing chains such as The Gap has been put down to their intelligent placing of the merchandise on touch tables that are at just the right height for us to run our hands over the garments as we stroll by.[39]

You might think that much of this sounds straightforward enough, but you would be surprised by how many stores get it wrong. For instance, when I organize tours with some of my clients in retail, we often come across clothing chains with their tables raised just a few inches off the ground. Someone really needs to tell them (the store managers, not my clients, that is) that the more difficult we find it to execute an action, the less we like whatever it is we touch or pick up. And while the effect in such cases might be small, it can nevertheless add up.[40] Shoppers would certainly seem to agree about the importance of touch, with 35 per cent of those questioned in one survey acknowledging that the feel of a mobile phone was more important than its look. Meanwhile, more than 80 per cent of those consumers quizzed in another survey said that they would choose a product that they could both see and feel over one that they were only able to look at.[41]

But this isn't something that we find only in clothing stores. Tables that enable the shopper to interact with, and experience, the products on offer are, of course, also a distinctive part of gadget stores. But have you ever noticed how all the MacBook Pro screens in the Apple store are tilted at exactly the same angle (70°, should you want to know)? This is far from the best viewing angle, so why, I ask you, would they do that? According to one commentator, it is so that shoppers will be tempted to adjust the screen in order to see better. In so doing, of course, they must physically touch the product itself.[42] Who said marketers weren't a sneaky bunch? Research by the Swedish economist Bertil Hultén has shown that the probability of shoppers touching the glassware in an IKEA store can be increased simply by turning the ambient lighting down,[43] and a dramatic increase in sales of those products (65 per cent) was reported when also adding a pleasant vanilla fragrance. (This figure is based on a sample of around 900 shoppers with sales monitored over two successive weekends.) Does something similar happen with the clothing in Abercrombie & Fitch, with all that dim nightclub-like lighting and intense scent?

Have you ever picked up a Bang & Olufsen* remote control? If so, you will know only too well its exceptionally and surprisingly weighty feel in the hand. The haptic experience just oozes quality. But what most people are not aware of is that the majority of that weight has been added for no purpose other than to convey this impression. That is, it serves no functional role whatsoever. Even more remarkable is the fact that the positive influence (or halo effect) of all that weight in your hand still

---

* For a number of years, this Danish company sold some of the sleekest, not to mention most expensive, luxury consumer electronics.

seems to work its magic even when you know why it is there. And, if you don't think you would fall prey to such a cheap trick, just ask yourself whether that expensive bottle of wine, lipstick or beauty cream that you just bought really needed to be quite so heavy? This is an example of what Louis Cheskin referred to as 'sensation transference'. The idea here is that the sensations we associate with what we feel seem to carry over to our experience of, and liking for, other product attributes as well. So, given that we all intuitively tend to associate weight with quality, we automatically believe that those products that come in heavier packages or containers are going to be better too. This presumably explains the correlation between the weight of the packaging and the price of the product. For instance, according to the results of one store audit carried out in Oxford with my colleague Betina Piqueras-Fiszman a few years ago, shoppers get an average of 8 grams more glass for every pound extra that they pay for a bottle of wine.[44]

## Tactile contamination

If you are anything like me, then you never pick the topmost newspaper or magazine from the pile, but rather select one from a little further down the stack. Prior to the Covid-19 pandemic this would have seemed like a pretty irrational thing to do, right? Presumably exactly the same copy can be found in each and every copy, as it were. But lots of people do it. Consumer researchers who have observed this somewhat unusual behaviour argue that it may reflect a subconscious fear of 'tactile contamination'.[45] That is, many of us simply do not like to buy products that have already been touched by someone else. So is there also a downside to all those 'Touch me' signs?

Think back to the last time you bought a new towel. Did you wash it before you started using it? I sincerely hope so, because the chances are that it had already been fondled by an average of six shoppers before you finally put it into your basket.[46] And, should you want to know just what to expect when lots of people touch the same surface, when researchers swabbed the screens at eight different UK branches of McDonald's they found faecal matter from several different people, not including the person who was trying to order their Big Mac and fries.[47] This, of course, is especially unfortunate given that fast food is almost always eaten with one's hands. Disgusting! Come to think of it, I think I'll go and wash my hands, not to mention those new towels, right now!

## *Can multisensory marketing really deliver a superadditive sales boost?*

Get the multisensory combination of atmospheric cues right and sales success will almost certainly follow. That, at least, is what some of the more effusive self-styled marketing gurus have been telling anyone who would listen for the last few years. That said, talk of a 1,200 per cent 'superadditive'* boost to sales from stimulating the shoppers' senses in just the right way is, I suspect, wishful thinking, despite the claims made in the marketing literature.† Turning to the results of

---

* 'Superadditivity' is a notion that originally emerged from the field of neurophysiology. The idea being that individually weakly effective sensory inputs might sometimes give rise to a response, in a neuron, in perception or in behaviour that is much greater than the sum of the responses to the individual inputs. You may remember it was mentioned briefly in the introductory chapter.
† I'll avoid naming them to save their blushes.

well-controlled academic studies that have been published in this area, the sales lift from combining sight, sound and smell to create a multisensory atmosphere normally tends to be much more modest – with a 15 per cent lift in sales being typical.[48] Does this mean, then, that multisensory atmospherics is simply not as powerful a notion as Kotler would have had us believe all those years ago? Perhaps, though, another way to think about it is that stores may already have taken on board many of the ideas around delivering the total experience. Thus, having already optimized their offering, it may just be that much harder to deliver further gains over and above what has already been achieved in the world of retail.[49]

That the senses interact really is not in any doubt these days. But which particular combination of atmospheric cues will be treated as congruent without involuntarily overloading the consumer's senses is a more challenging question to answer. In one study, for instance, participants were presented with all nine possible combinations of no scent, low-arousing scent (lavender) and high-arousing scent (grapefruit) together with no music, low-tempo music or high-tempo music. If the scent and music were congruent in terms of their arousal potential, then customers rated the store environment (a gift shop) more positively. Shoppers also exhibited higher levels of approach and impulse-buying behaviour, and experienced enhanced satisfaction.[50]

One of the only studies to have been conducted to date in the mall setting illustrates the potential problems faced by marketers working in the field of multisensory atmospherics. Morrin and Chebat looked at the spending on unplanned purchases by nearly 800 shoppers in a North American mall. Their results suggested that such sales could be increased by as much as 50 per cent simply by playing slow-tempo music. By contrast, releasing a citrus fragrance led to a small (and non-significant) decline in sales. However, when the music and

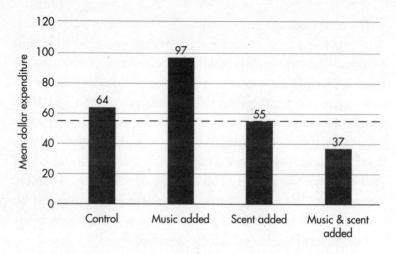

An example of sub-additivity in the marketplace.

fragrance were presented at the same time, sales in the mall dropped significantly. What went 'wrong' in this case is hard to say without knowing more. But one possibility is that the musical and olfactory stimuli may simply have been incongruent along some dimension (such as perhaps their arousal value). Indeed, it is easy to imagine how slow-tempo music would have relaxed the shoppers while the citrus scent may well have aroused them.[51] As such, combining an arousing scent with relaxing music may have left shoppers confused. We find it hard to process incongruent signals.*

While introducing more sensory cues into a store atmosphere is a good idea in terms of increasing the number of sensory touch points, it can also increase the risk of sensory overload, as illustrated by a study in which 800 people had to

---

* The lack of processing fluency that is associated with mismatching inputs is usually negatively valenced – meaning we don't like it, and hence it doesn't do much for sales, as Morrin and Chebat's data would seem to imply.

imagine themselves browsing in a store. In the imagined store, there was fast or slow music, the scent of lavender or grapefruit, and a red or blue colour scheme. Just as long as any combination of two atmospheric cues was congruent the outcome was positive. However, as soon as three congruent stimuli were introduced into the imagined scenario, negative effects started to appear. Homburg et al. attributed this to the level of arousal simply being too high, possibly reflecting some sort of sensory overload. What would happen in a real store setting remains to be seen, of course.[52]

## *Time to turn the lights up, and the music down?*

The very real danger, then, when thinking about stimulating shoppers' senses, one sense at a time, is that a scent that, in isolation, might well help to alert or attract the shopper might simply prove too much if combined with a high-tempo loud musical selection, or else some especially bright store lighting. Combining atmospheric cues can all too easily lead to sensory overload.[53] This is precisely what many parents complain about when standing outside a branch of Hollister or Abercrombie & Fitch, say, their senses assaulted by the loud dance music and that prominent scent while their children are merrily shopping away in the dark nightclub-like interior. *How can anyone stand that racket?* you find yourself thinking. Undoubtedly, part of the aim is to dissuade uncool old wrinklies from entering. Or, as Mike Jeffries, then CEO of Abercrombie & Fitch put it back in 2014, 'We want to market to cool, good-looking people. We don't market to anyone other than that.'[54]

At the same time, however, that distinctive energetic upbeat music may also be helping to enhance sales. After all, it is well known in the world of food and drink that playing loud, fast music can boost sales by as much as 30 per cent. Why wouldn't the same be true for clothing? One does, though, have to feel sorry for the sales staff in clothing stores who, like an increasing number of those working in the restaurant setting, are exposed to dangerously high levels of music on a regular basis.[55]

Both scent and sound are used to drive sales of clothing as well as other intrinsically odourless products (e.g. books and magazines). That said, there is currently no simple means of predicting, and hence avoiding, the multisensory mismatch or overload that can sometimes result when all of the shopper's senses are stimulated at the same time. It is for this reason that I recommend to my clients that they should consider conducting the research themselves in their own establishments. That way, they can determine what works best in terms of driving sales in their own customer base, rather than simply relying on research that was inevitably conducted in another time and place.

Are you one of those individuals who loves the Lush experience and doesn't know what all the complaints are about as far as A&F is concerned? If so, you might well be a 'sensory junkie'. This is a term used to describe those shoppers who crave multisensory stimulation.[56] Some researchers have tried to distinguish between impulsive and contemplative shoppers. According to Morrin and Chebat, for example, the former might be more influenced by background music, whereas the latter might be more influenced by scent. For those of us who are not sensory junkies, who cross the street as soon as we sniff a Lush store and who have perhaps never ventured into an A&F or Hollister store, then help is at hand. A few years ago, Selfridges in London introduced a chill-out room so that weary shoppers could recover from all that hectic multisensory stimulation out on the sales floor. One other suggestion is simply to put a pair of earplugs in before you next go shopping to attenuate the din – or perhaps use the NozNoz device from page 31.[57]

## Tasting the future

Enhancing the in-store tasting experience is currently also of great interest to many marketing agencies and food and beverage brands. For instance, in 2017 I consulted on an innovative approach in this area in which we tried to reinvent the multisensory tasting experience in the context of the supermarket. Shoppers in the UK were approached in the drinks aisle at branches of Tesco and invited to put on a VR headset. They got to sample three beers, Guinness Draught, Hop House 13 Lager and West Indies Porter, while being led through the branded tasting experience by the voice of the charming Guinness Master Brewer, Peter Simpson. As they tasted each drink, a specially created 360-degree audio-visual display was piped to their headset, with each element designed to match the drink that the shopper was tasting.

Crucially, the sounds, colours, shapes and patterns of movement had all been designed specifically to match the flavour profiles of the drinks that were being sampled, and so hopefully enhance the taster's flavour experience.[58] The campaign was enthusiastically received by shoppers, and illustrates one of the many ways in which technology is being brought to retail.[59] Sensehacking the multisensory tasting experience using the latest in digital technology – and just about as far away from the traditional taste test where consumers are given a thimbleful of some drink or other from a cheap light plastic cup. Don't those promoting such events realize that everything tastes better in a heavier glass?[60]

## Multisensory shopping online

At the turn of the century, it was often said that consumers would never be convinced to buy clothes online. This prediction now seems quaintly outdated, to say the least.[61] At the same time, however, when we do buy online, we get no sense of the feel or fit of clothing or shoes. No wonder, then, that returns have reached epidemic levels. According to a 2017 report in the *Financial Times*, dealing with returns is costing retailers £60 billion a year in the UK. In the US, it has been estimated that return deliveries will cost businesses $550 billion in 2020, 75.2 per cent *more* than in 2016. And while people do, of course, return their purchases to bricks-and-mortar stores, return rates for online purchases are just much higher.[62] The phenomenal rise of personalization in retail may help to address this problem in some small way. After all, just think about how much harder it is going to be to return those personalized Nike Air trainers, or that exclusive Louis Vuitton handbag, once it has your initials emblazoned all over it.

Currently, the challenge for many digital marketers is that they only have access to one, or at most two, of our senses – sight and sometimes also sound. These are the higher, rational senses. Of course, the technologists and retail futurologists have long been promising that our computers and, more recently, our smartphones would very soon let us feel the softness of that new cashmere sweater or those silky-smooth pyjamas.[63] It has also been confidently suggested that advances in technology would soon let us smell the perfume that we are thinking of buying, or even taste that pizza that we are hesitating over whether or not to order. You do not need me to tell you that this simply has not happened. What is more, there would seem to be little likelihood of the situation changing any time soon.

And should you want to know what became of all those companies who were selling us these pipe dreams, the majority went under years ago, taking all that over-eager venture capital funding with them. Or, as one headline put it, referring to a failed start-up that promised to deliver product scents for those shopping online, 'What does $20 million burning smell like? Just ask DigiScents!'[64]

This is a real shame, given that there are products that people seem simply unwilling to buy online. If you do not believe me, just ask yourself whether you would buy a new perfume or aftershave, say, without having smelled it first? I certainly don't think that I would. Given that technological solutions have yet to emerge, this will likely limit the growth of online retail for at least some products, given that it is impossible to deliver the full multisensory experience with the technology that is currently foreseeable. Instead, we will probably see much more of synaesthetic marketing (as in the Guinness example just mentioned), as well as possibly the delivery of some more extraordinary experiences, as we will see with our final example.

## On the future of online marketing

In a project that I believe hints at where digital marketing may be going in the future my colleagues and I worked closely with Glenmorangie whisky distillers in 2018. There, we established the optimal sensory triggers for the autonomous sensory meridian response (ASMR) that linked to Scottish themes appropriate to the whisky. ASMR is the name given to the relaxing tingle down the back of the neck that many people experience when listening to someone whispering or crinkling paper. Some of the key triggers identified by our research, which involved interviewing the large online community of regular ASMR-ers, included slow-paced close-ups with realistic sound and an absence of background music. High-pitched sounds and texture also turned out to be important. These and other stimuli that our interviewees recommended including were then used as scientific inspiration by three video artists – Thomas Traum, Julie Weitz and Studio de Crécy – to produce films designed to evoke the whisky's 'terroir, creation and character', one video for each of the three different expressions of Glenmorangie: The Original, Lasanta and Signet.

On launching the activation, consumers were invited to pour themselves a glass of their preferred expression of the whisky, plug in a pair of over-ear headphones and then access the relevant video content online. Hinting at the appetite for such digitally mediated content, this turned out to be the most successful campaign in the company's history of online marketing. It is not often, after all, that a drink gives you a relaxing tingling shiver down your spine.[65]

The Glenmorangie campaign, like the Guinness example, shows how the latest in digital experience, together with the emerging knowledge of perception, can be used to hack the multisensory experience in the home online, or in store. Figuring out how such experiences can be delivered, enhanced and propagated digitally is one of the most intriguing challenges for sensehacking in the years to come.

## *Shop 'til you drop*

So, no matter whether you are a sensory junkie or not, you will perhaps now understand a little better why it is that we all find it so difficult to stop shopping. More than anything, the science of sensehacking explains why it is that so many of us seem to shop 'til we drop. From sights to sounds, smells to touch and even temperature, all have been deliberately designed and/or are carefully controlled to create just the right multisensory atmosphere. Is it any wonder, then, that we end up staying longer and picking up more than perhaps we should?

But whatever you come home with, just remember to wash your hands afterwards if you want to avoid the dangers of tactile contamination. Never mind coronavirus – just remember what they found on those fast-food touch screens!

# 8. Healthcare

Let me start by asking you a question: Would you mind if the surgeon who was operating on you listened to music while they worked? While this is not something that most of us have probably given much thought to, music is played in the majority of operating theatres. According to a 2014 study published in the *British Medical Journal*, music, most often classical, can be heard somewhere between 62 and 72 per cent of the time in surgery.[1] It turns out that just like in the restaurant kitchen, listening to music can help relieve the boredom associated with standard operating procedures: hip replacements and the like in the one case, brunoising a bucketful of potatoes or carrots in the other. Indeed, according to a recent review co-authored by the appropriately named British surgeon Roger Kneebone and modernist chef Jozef Youssef, there are actually far more similarities between what goes on in a professional restaurant kitchen and in the operating theatre than you might think.

Given the large body of research suggesting that our performance tends to be entrained to the musical beat, one might naturally wonder whether surgeons would also work faster when listening to higher-tempo music. But it is not just the speed of the music that matters, getting the genre right is important too. I do not know about you but I don't think that I would feel comfortable with my surgeon listening to death metal while they were working on me, nor, for that matter, 'Another One Bites the Dust' by Queen, or even 'Everybody Hurts' by R.EM. And, for those who are addicted to a bit of nip-and-tuck, 'Scar Tissue' by Red Hot Chili Peppers might seem a little insensitive were it to make its way onto your plastic surgeon's playlist.*

While it can be ethically challenging to conduct appropriately controlled studies in patient populations, it turns out that trainee plastic surgeons working on pigs' feet apply their surgical wound closures significantly faster when listening to their preferred music than when working in silence.[2] In this case, the presence of background music resulted in an 8–10 per cent reduction in wound repair time, along with the closures themselves being rated as higher in quality by their peers. Meanwhile, the fifty male surgeons in a study by Allen and Blascovich exhibited both a smaller increase in heart rate and improved performance when tackling a stressful lab task if they were allowed to listen to their preferred choice of music rather than working in silence or being forced to listen to the music provided by the experimenters.[3] Given the high cost of surgical provision, one could certainly imagine a recommendation coming from the hospital accountants charged with trying to cut costs to turn up the musical tempo. After all, operating theatre costs exceeded $60 per minute in North American hospitals back in 2005, which meant that savings of more than $100,000 could be made by reducing operating time by approximately seven minutes for just 250 cases.[4] The figure will be a great deal higher today. Every little helps, as they say.

Do you remember a battery-powered children's game called Operation? Players would take turns trying to remove the various organs and bones out of the holes of the unlucky patient pictured on the board using tweezers. Sooner or later, someone would accidentally touch the sides, thus short-circuiting the

---

* Go on then, I challenge you – surely, you can think of a few more inappropriate tracks to include on the surgical playlist, you know, the ones that no one would want to hear?

patient, fondly known as Cavity Sam, causing his nose to start flashing red and a buzzer to sound. In 2016, 352 members of the public were invited to play this game at the Imperial Festival in London while listening to one of three soundtracks. The performance of men, but not women, was worse when listening to Australian rock music than to the pre-recorded sounds of an operating theatre. They were significantly slower and made more mistakes, suggesting that they found it harder to concentrate. By contrast, those who listened to Mozart perceived the music to be less distracting, although surprisingly that did not translate into a quicker operation or fewer mistakes, contrary to what the literature on the 'Mozart effect' might suggest.[5] I'll leave it up to you to judge the implications the results of such decidedly light-hearted research might have for the music you'd rather that your surgical team listened to.

In another, more serious study, professional anaesthetists reported that they found reggae and pop music particularly distracting.[6] But no matter what the music, it is unlikely that all of those working in the operating theatre are necessarily going to be happy with the selection, and, indeed, tensions have been known to result when a surgeon chooses an eclectic playlist.[7] So, the next time you go in for some routine surgery, it might pay you to find out what the surgeon plans to listen to while you are under the knife. It really can make a difference to how they, not to mention their colleagues, perform. And this is just one example of how the senses can be hacked to influence healthcare outcomes.

## *Why hospitals are starting to look like high-end hotels*

From our first breath until our last, and periodically in between, we all rely on the healthcare system to support us. It provides assistance to facilitate normal development while trying to deal with those problems that inevitably arise along the way. Traditionally, medical solutions (as opposed to sensehacking) have always been at the forefront of treatment, no matter what stage of life we are at. Intriguingly, though, back in 1974, the legendary marketer Philip Kotler was already writing presciently about applying his atmospherics approach to retail to the design of the psychiatrist's office.[8] More recently, this has been followed by the steady growth of interest in applying ideas around the 'experience economy' to healthcare provision. Such thinking tends to be a little more common in the private healthcare system, at least in the UK. That said, hospitals, doctors, dentists, care homes and even plastic surgeons around the world, no matter whether publicly or privately funded, are increasingly recognizing the value of adopting a more multisensory approach to healthcare. Indeed, a growing number of them now see it as an essential component of the service, or experience, that they provide. In the United States, a portion of some physicians' final salaries are determined by the patient satisfaction ratings they receive.[9] Such satisfaction-related pay is already a feature of more than 40 per cent of physicians' packages in the US, with that figure set to rise.*

This shift in emphasis has been driven, at least in part, by increased competition amongst private healthcare providers in certain parts of the world. For, when it is no longer possible to differentiate a medical offering to potential customers or stakeholders in a meaningful way – with better scanners or newer facilities, for example – thoughts increasingly turn to the quality of the patient experience. This experiential transformation is

---

* Is there a link to the current opioid crisis in the States, one wonders. After all, denying your patient the painkillers that they crave is one sure way to lower your ratings.

seen in those private hospitals that look more like high-end hotels than conventional healthcare facilities; or, as the headline of one press article put it early in 2019, 'The Mayo Clinic: looks like a hotel, but this is the best hospital in the world.'[10] The same article goes on to note that people come from around the world for the quality of the patient experience. But it is important to note that this is more than merely a matter of sprucing up appearances. The quality of the patient-rated experience really does correlate with healthcare outcomes.[11] As such, focusing on patient satisfaction may well turn out to be an effective route both to improving the quality of medical care and, at the same time, potentially also to delivering a cost saving in the long term. This is important, given that anything that improves treatment outcomes is likely going to reduce costs too, which ought to keep the hospital accountants happy.

For example, according to the results of a 2016 study analysing risk-adjusted data from more than 3,000 US hospitals released by the Centers for Medicare & Medicaid Services (CMS), better patient experience was associated with favourable clinical outcomes. According to the study's authors, 'a higher number of stars for patient experience had a statistically significant association with lower rates of many in-hospital complications. A higher patient experience star rating also had a statistically significant association with lower rates of unplanned readmissions to the hospital within 30 days.'[12] Meanwhile, the results of another study, published the following year, involving almost 20,000 observations from 3,767 hospitals in the US over the six years from 2007 to 2012 concluded that positive patient experiences were associated with increased profitability. By contrast, a negative patient experience was found to be even more strongly associated with decreased profitability.[13]

In an earlier chapter, we took a close look at the beneficial effects of exposure to nature on our well-being. Nature has long been incorporated into healthcare provision via the healing gardens that are tucked away in any number of hospitals and care facilities. In fact, the earliest Western hospitals would have been dependent on herbs, plants and perhaps a cloistered garden for healing. For many years now, such tranquil spaces have helped to provide patients and their families with comfort and solace in their hour of need. In one study, spending an hour in the garden outside the care home led to a significant improvement in the ability of very elderly patients to concentrate when compared with spending the same amount of time in their favourite room.[14] According to an article in *Scientific American*, the most soothing gardens engage multiple senses: 'Gardens that can be seen, touched, smelled and listened to soothe best.' The article also highlights that while they were 'dismissed as peripheral to medical treatment for much of the twentieth century, gardens are back in style, now featured in the design of most new hospitals'.[15]

During the Crimean War of 1853–6, Florence Nightingale was already advocating the beneficial effects of quietude, not to mention natural lighting, on the recovery of the patients she was caring for. As we will see again later, this truly inspirational nurse was certainly way ahead of her time as far as sensehacking healthcare on the wards is concerned. However, it was Ulrich's seminal finding of the positive impact on the recovery of a small group of surgical patients given a hospital room with a view, published in 1984, that really kick-started the contemporary interest in 'salutogenesis'. The term was first introduced by Aaron Antonovsky in 1979 to refer to an approach to medical practice that emphasizes those factors serving to support human health and well-being, rather than focusing solely on those responsible for causing disease (pathogenesis). A salutogenic environment can be characterized as one in which the multisensory features of the atmosphere have been orchestrated to aid patients' recovery.[16]

## Tastes healthy

While adopting a salutogenic approach might sound simple enough in theory, it is surprising just how often the sensory aspects of the healthcare environment/experience currently appear to be having the opposite effect on patient well-being. Think only of the background noise in hospital wards that prevents so many patients from getting a good night's sleep.[17] Or all the unappetizing hospital food that gets returned to the kitchens uneaten every day. According to the results of one survey, an astonishing 70 per cent of all food served to patients in National Health Service hospitals in the UK is returned untouched.[18] These are just two of the most obvious, and widely discussed, examples of poor healthcare provision, perhaps attributable to an overemphasis on pathogenesis rather than salutogenesis. One key objective for those wanting to sensehack our health involves identifying those negative sensory factors that may be deleteriously impacting healthcare provision and trying to address them. Taking the two examples just mentioned, this would involve trying to reduce ambient noise levels and improve the quality of the food.*[19] But just because such solutions are in some sense obvious doesn't mean that they can't have profoundly beneficial outcomes on patient well-being.

For example, in one two-year NHS trial at six hospital trusts in the UK, elderly patients' chances of dying in hospital following a hip fracture were halved, yes, that's right, halved (down from 11 per cent to 5.5 per cent), simply by giving them an extra meal each day and encouraging them to eat it. On

---

* At the same time as counteracting the negatives, though, there is also growing interest in proactively trying to offer innovative sensory, and increasingly multisensory, interventions that, in some cases, are even tailored to the individual, such as personalized nutrition.

seeing the results of the study, chief orthopaedic surgeon Dominic Inman was moved to suggest, 'If you look upon food as a very, very cheap drug, that's extremely powerful.'[20] That said, patients in the trial were asked in the morning what they wanted to eat by a nutritionist, who then sat with them at mealtimes to make sure they finished their meal. The cost implications of providing such a personalized service are, though, likely to soon raise their ugly head. Meanwhile, over in the States, a study by my friend Dr Claudia Campos and her colleagues has highlighted the beneficial effects of serving meals in accordance with the Dietary Approaches to Stop Hypertension (DASH) diet to a multiethnic cohort of patients. These researchers documented a significant long-term reduction in the risk of heart disease amongst those who followed the dietary guidelines most closely.[21]

It is therefore interesting to see how a range of low-cost sensory interventions have, over the years, provided an effective means of increasing how much patients eat. For example, patients with Alzheimer's and/or dementia can struggle to distinguish the food from the plate. Pale hospital food – think mashed potato, creamy sauces, chicken and the majority of fish fillets – simply do not stand out on white crockery. Something as simple as using high-contrast brightly coloured red or blue crockery and cutlery and beakers has been shown to increase consumption by as much as 30 per cent, in both hospitals and long-term care homes, and so it is easy to understand why a number of start-ups have emerged in recent years offering such visually enhanced crockery to the general public.[22]

Another relatively simple low-cost sense hack at mealtimes is to use music or ambient soundscapes to help relax those patients who might otherwise be too agitated to eat. This is a particularly common problem amongst many psychiatric patients as well as a growing number of those suffering from

Alzheimer's/dementia. Intriguingly, back in the 1970s, a number of North American psychiatric hospitals were already playing a tape entitled *Sea gulls . . . Music for rest and relaxation* for just this reason.[23] This was long before Heston Blumenthal's world-famous *Sound of the Sea* dish, which consists of a plate of sashimi brought to the table together with an MP3 player in a conch shell playing seagull cries and the sound of waves lapping gently on the shore. The aim, in both cases, was to sensehack the experience, although the hope in the restaurant was not to calm agitated diners but to trigger positively valenced nostalgia in them.

A few years ago I was lucky enough to consult on the development of an award-winning aromatic scent-delivery system: a food-scent alarm clock, if you will. The device, called 'Ode' (after the scent-expert involved in the project, whose moniker was 'Odette de Toilette'), released hunger-inducing food aromas three times a day in the homes of those who might otherwise forget to eat. The six scents developed for the launch included fresh orange juice, cherry Bakewell tart, homemade curry, pink grapefruit, beef casserole and Black Forest gateau (chosen to be representative of familiar food aromas to those in the target age group). The admirable aim of this project was to help those early stage Alzheimer's/dementia patients who might otherwise need to be hospitalized due to undernutrition to be able to live at home for a little longer than would otherwise be the case. The results of a ten-week small-scale pilot study involving fifty people with dementia, along with their families, who used the device revealed that more than half of them did indeed end up maintaining their weight, or else showing a slight increase, as compared to the weight loss that is so often seen in this group. This, then, is just one of the low-cost sense hacks that might help nudge those who need it to eat more regularly.[24]

## Looks healthy: the art and illusion of healthcare

As in the workplace, a room with a view of nature appears to benefit recovering surgical patients. This isn't possible for everyone in hospital, so we need some sensehacking alternatives.* Potted plants might be a start since they can help reduce stress while also making the clinical environment appear a little less intimidating.[25] Putting art on the walls can also help improve patient outcomes. Indeed, a 2006 report from the UK Department of Health Working Group on Arts and Health concluded that the arts have 'a clear contribution to make and offer major opportunities in the delivery of better health, wellbeing and improved experience for patients, service users and staff alike'.[26] Let me illustrate with an example. Many people exhibit elevated blood pressure in a medical setting, something known as 'white-coat hypertension'. Intriguingly, this phenomenon can be reduced simply by mounting landscape photographs on the walls of the examination room.[27] Meanwhile, pain levels are reduced, and clinical outcomes improved, in patients who are encouraged to engage with art.[28]

It is the presence of all that art that is presumably one of the main reasons why more and more hospitals have started to look like 'high-end hotels'. Note, though, how such an easy fix is one that presumably any public hospital, no matter how tight its budget, might emulate, even if they can't afford to compete with the fabulous art collection at the Mayo Clinic, which includes works by Rodin, Andy Warhol, Dale Chihuly and many others. Once again, it was Florence Nightingale

---

* The absence of natural lighting in the ICU is also associated with an increased risk of delirium in older patients, especially those with dementia, a phenomenon that is referred to as 'sundowning'.

who intuitively picked up on the healing power of art, writing in 1860:

> The effect . . . of beautiful objects, of variety of objects and especially brilliance of colour is hardly at all appreciated. Little as we know about the way in which we are affected by form, by colour and light, we do know this, that they have an actual physical effect. Variety of form and brilliancy of colour in the objects presented to patients are actual means of recovery.[29]

Art on the walls of a hospital not only helps to set our expectations about healthcare outcomes but arguably also comprises part of the treatment it offers. Chromatherapy, or colour therapy, is an approach to hacking the senses that became popular in the closing decades of the nineteenth century. According to its proponents, patients suffering from rheumatism, inflammation, nerves, schizophrenia, or other forms of mental illness could be treated simply by bathing them in light of particular colours. According to one of the proponents, Edwin Babbitt, red light was prescribed for physical exhaustion and chronic rheumatism, yellow as a laxative and for bronchial difficulties, blue for inflammatory conditions, and so on.[30] While chromatherapy is nowadays considered to be little more than pseudoscience by mainstream healthcare professionals, it is worth remembering that colour and light clearly do exert a significant impact on various aspects of our social, cognitive and emotional well-being. Just think back to how the blue light of dawn was shown to keep us alert, while painting police holding cells bubblegum pink supposedly helps calm agitated prisoners.* While blood-red walls may help to make the landscape paintings look good in the picture gallery, it's hard to believe that the same paint colour

---

\* The same Baker-Miller pink as discussed in Chapter 2.

would be anything other than very disturbing if it were to find its way onto the walls of a hospital.

Clearly, then, colour and lighting play an important role in hospital design, as does the art on the walls.[31] Indeed, the colours in many of the most highly regarded facilities are chosen for their psychological effect, designed to be appropriate to the situation. For example, at the Mayo Clinic, 'Even the colours of the walls are carefully chosen to encourage certain moods – so soft shades of blues, greens and violets are used in cancer diagnosis areas to promote calmness and lower stress levels. Consulting rooms are painted blue, which researchers from the clinic have found to be better for building trust.'[32]

While we are on the topic of the psychological use of colour in the hospital setting, have you ever stopped to ask yourself why it is that surgical scrubs and ward curtains are so often green? This sense hack helps to reduce any visual after effects for the theatre staff who spend so much of their time looking at their patients' bloody insides. After staring intently at something for a long time, we tend to see a negative afterimage of whatever we were concentrating on when we look away. And since green is the opponent colour to red, the afterimage of your insides that the surgeon sees when they look away will be tinted green. The sense hack, then, is that such afterimages will be a little less perceptible, and hence less distracting, when looking at a green surface.

What is perhaps even more impressive in terms of using visual cues to promote healthcare outcomes is how some multisensory illusions, adapted from the psychology lab, have also shown promise in terms of helping to ameliorate the intractable chronic pain suffered by several groups of patients for whom traditional approaches to pain relief offer no respite. Sensehacking has already shown great promise in terms of helping patients who are suffering from phantom-limb pain or complex regional pain syndrome (CRPS). One much-publicized psychological intervention concerns the mirror box. This visual, or – more accurately – multisensory, illusion has been used to help reduce the phantom-limb pain experienced by the aforementioned patients, many of whom suffer with excruciating pain in a limb that is no longer there.[33] These patients typically had a limb surgically removed as the result of an accident that led to great pain. If the patient's last memory of the affected limb was of it being painful, then, in some unfortunate cases, that painful sensory trace remains.* So, despite the fact that the affected body part is no longer there, in the absence of any discrepant sensorimotor feedback, it can be hard for the patient's brain to update this last impression. Instead, the patient may well be stuck with a painful phantom limb that they cannot move, and which often feels like it is cramping.

The idea behind the mirror box is to give the patient the illusion that their missing limb has been replaced, using the mirror reflection of their intact limb. As the patient moves their unaffected limb, they see what appears to be their missing limb moving in the mirror. This discrepant feedback causes the painful phantom appendage to shrink, thus reducing the associated pain. That said, subsequent research has questioned whether the benefits of mirror therapy, such as they are, reflect the direct result of hacking the patient's senses – that is, by making it look as though their absent limb is still there. According to an alternative account for why this approach works that I put forward together with lead author Professor Lorimer Moseley when he worked in Oxford a few years ago, it may simply be the motor imagery that is evoked by getting the patient to think

* Nowadays, enlightened surgeons try to make sure that the limb has been anaesthetized before it is removed.

An example showing how a mirror box can be used to trick a phantom limb patient into seeing their missing limb.

about moving their phantom limb that is really doing the work here.[34]

An equally dramatic means of helping to reduce pain in those with CRPS was discovered in my Oxford base. CRPS patients often present after having suffered damage, such as a fractured wrist following an accident. Initially, they appear to be recovering perfectly normally; however, after six months or so, these patients start to develop excruciating pain in the affected limb. What is more, the limb often shows signs of swelling, and doctors have noticed anecdotally that it tends to feel cooler than the unaffected limb – this being one of the standard bedside tests for the condition. Once again, painkillers typically do not work, and often the pain is so unbearable that the patients may even request, or rather beg for, its surgical removal. The truly horrendous impact that CRPS can have was brought home to me when my former supervisor Professor Jon Driver, the person who is primarily responsible for my ending up in this field of research, threw himself from a motorway bridge a few years ago. He had suffered from severe CRPS in his leg after having been knocked off his moped.

Together with lead researcher Lorimer Moseley and a group of ten long-suffering CRPS patients, we made their affected limbs look smaller by having them look through a minifying lens – think of what you see when you look through a pair of binoculars backwards. Amazingly, this simple sense hack significantly reduced subjective pain ratings in a matter of minutes. What is more, an objectively measurable index of the condition, namely the swelling in the affected limb, was also reduced after a few minutes of minification.[35] Such results are consistent with other research that we have published showing that using a joke-shop rubber hand to convince you that one of your arms has been replaced also results in physiological changes, such as a rapid drop in the temperature of the affected, or replaced, limb.[36]

Given that some in the medical community were initially sceptical that hacking the senses using visual illusions could elicit such rapid changes, it is gratifying, not to mention reassuring, to see that a number of other research groups have since confirmed our findings.[37] So, while longer-term clinical follow-ups are undoubtedly still needed in order to establish the robustness and longevity of such solutions, it should be clear that both art and illusion hold great promise in terms of hacking the senses in healthcare. In many ways, though, the ultimate test in this regard is whether the patients themselves find the benefits to be worth the effort.

## *Healthy hearing*

Let us return to the auditory aspect of healthcare, as this is where most problems are currently found, and where some of the most intriguing solutions reside. Once again, Florence Nightingale was on the money more than 150 years ago when she wrote, 'Unnecessary noise is the most cruel absence of care which can be inflicted either on sick or on well.'[38] The sonic element of healthcare may become all the more salient in the patient's experience precisely because the visual environment in many hospitals is typically so dull and boring.[39] As anyone who has been in a hospital ward or intensive care unit knows only too well, they tend to be exceptionally noisy places, due, in large part, to the many alerts, alarms and other digital warning signals that can be heard on wards these days.[40]

While such high noise levels are bad enough during daylight hours, at night they can become especially disturbing. Noise levels have been increasing exponentially in hospitals in recent decades. According to guidelines from the World Health Organization, noise levels on hospital wards should not exceed 35dBA during the day and 30dBA at night.*[41] For reference, 30dBA is about the same loudness as a whisper. It is shocking, therefore, to find peak levels in hospitals of more than 80dBA, equivalent to the noise made by a chainsaw.[42] In an observational study in the UK, noise levels in five ICUs were 60dBA during the daytime, with peaks above 100dBA recorded every two to three minutes (equivalent to a noisy motorcycle or handheld drill). Although things quietened down at night, peak sounds above 85dBA were still identified up to sixteen times an hour.

No wonder so many patients complain that they simply cannot sleep in hospital. Being interrupted by noise every six minutes during the night while critically ill must surely be having a negative impact on healthcare outcomes.[43] One poignant case involved a former chief scientific advisor to the UK government, Sir

---

* 'A' weighting for dB levels (i.e. dBA) indicates that the sound pressure levels have been weighted to approximate how we hear.

David MacKay. As a patient, he was driven to tears by the relentless noise on his hospital ward. As he put it in a blog posted the day before he died, 'The ward is always full of lights going on and off, doors opening, special mechanical beds that make fizzing electrical noises and clunks for hours on end.'*[44]

The potentially damaging effects of excessive noise are not just problematic for patients. The operating staff themselves also suffer, with orthopaedic procedures known to be especially bad. Peak sound levels of 120dB or more are routinely recorded once the electric sawing, drilling and hammering begin, with knee replacement surgery and neurosurgery known to be particularly noisy. For comparison, military jet afterburners produce 130dB during take-off. Such noise levels are potentially damaging to the hearing of those working in the operating theatre, who are regularly exposed to them. They may, however, be even more damaging to the patients themselves, this despite the fact that they are hopefully oblivious of the loud noises around them. The anaesthetic routinely administered during surgery tends to paralyse the stapedius muscle, which normally protects our ears from damage by attenuating the response to loud noise.[45]

Another loud noise that most of us unfortunately know only too well is the sound of the dentist's drill.[46] Surely if someone could simply eliminate that high-pitched whining noise, or maybe offer the patient some noise-cancelling headphones, and/ or allow them to distract themselves by watching a movie, the whole process of getting a filling would be less traumatic. Changing the sound for something a little less unpleasant might

---

* You don't need to be a genius to realize that patients who are denied a good night's sleep are probably not going to recover as rapidly as those who can use the night to recuperate. Notice here also how slower recovery presumably also correlates with increased costs. Of course, much the same argument can be made for those denied adequate, and here I mean only palatable, nutrition.

help too. Indeed, research led by Tasha Stanton of the University of Adelaide, a visitor to my lab in Oxford in 2019, reported findings consistent with just such a suggestion. Chronic back pain patients exhibited enhanced flexibility if a smooth pleasant sound, rather than the sound of a creaking door, was synchronized with the bending of their stiff back.[47] Notice how the sound is irrelevant to pain and movement, and yet synchronizing pleasant sounds with painful procedures improves the situation somewhat, no matter whether you have a bad back or a crumbling molar.

If I had my way, the other part of the solution to making a trip to the dentist's less painful would be to invent some means of allowing me to distract my attention from the oral cavity without my mouth inadvertently closing while the dentist is working on it. Attending to, or concentrating on, pain, or the location where it hurts, makes the experience more unpleasant. By contrast, distracting one's attention from the painful stimulus, or location, be it with the sights and sounds of nature or immersion in some other world, has been shown to help reduce its saliency.[48] Indeed, this is a large part of the reason that VR headsets are now being offered to patients in the hope of distracting them from incredibly painful procedures where they remain conscious, such as, for example, wound dressing.[49]

Recently, a Welsh hospital has also started offering VR distraction for another extremely painful occurrence: childbirth.[50] Irene Tracey, the Nuffield Professor of Anaesthetic Science at Oxford University who was dubbed by the media the 'Queen of Pain', defined the 'ultimate pain' (10 on the Montreal Pain Scale)* by saying, 'I've been through childbirth three times, and my ten

---

* Otherwise known as the McGill Pain Questionnaire, this is a standardized self-report scale used widely to capture the quality and intensity of pain that someone is feeling.

is a very different ten from before I had kids. I've got a whole new calibration on that scale.'[51] But beyond dealing with the excessive noise levels, be it on the wards, in the ICU, operating theatre, or dentist's surgery, how else does what we hear affect us in the context of healthcare?

## *Music therapy*

Earlier we saw how music can help surgeons do their job, but can it help patients too? That is, can music be used to soothe the pain? While the claim that music acts as a painkiller may well sound dubious to some, the evidence from many studies now shows that indeed it can. In fact, the beneficial effects of music have been convincingly demonstrated at all stages of healthcare provision. Not only during surgery itself, as we have seen already, and where the benefits are primarily experienced by the staff, but during all stages of the patient's perioperative care. Music is, for example, used to help relax patients in advance of medical interventions while also helping to distract them from a variety of painful procedures. Music has been used to calm women waiting for the results of a breast biopsy as well as provide relief to patients on mechanical ventilation.[52] Conrad and colleagues found that music helps decrease the dosage of sedative needed to achieve a desired level of calm in the patient.[53] Several other studies have confirmed the sedative and analgesic-sparing properties of music.[54] In total, many hundreds of experiments, together with a large number of Cochrane Reviews,* have now been published highlighting the benefits of music in healthcare.

---

* These independent reviews from the Cochrane Policy Institute are widely considered to be the gold standard as far as judging the empirical evidence (both the significant findings as well as the null results) on a given healthcare topic is concerned.

Music can be used not only to help reduce a patient's anxiety and stress but also to deal with pain, not to mention potentially reducing recovery times in the process.[55] Crucially, however, a sound business case – if you'll excuse the pun – needs to be made for what can seem, at least to some, like non-essential elements of the service provision – for example, live music being performed in one Irish hospital.[56] Healthcare is primarily focused on the treatment and prevention (or vice versa) of illness, not on entertainment. Hence, being able to demonstrate a link between the sensehacking of some aspect of the healthcare experience and reduced costs and/or improved outcomes will be key to its continued uptake.

Looking to the future, it won't be too long before someone comes up with an app that helps select music to match the various healthcare situations that one might find oneself in. And, beyond that, you really have to ask why, if music is so important to healthcare, we have mostly been relying on music that was composed for another purpose, or situation, to do the job.* Why not specifically compose music for healthcare? One innovative example along just these lines was created a few years ago by the legendary Brian Eno, who composed music specifically for those dealing with bad news, creating a healing ambient soundscape for the patients at Montefiore Hospital in Sussex.[57]

## Healing hands

Many of us are suffering from touch hunger. Stimulating the skin by stroking, caressing or massage has a profoundly beneficial effect

---

\* One of the few exceptions here would seem to be music traditionally associated with funerals and grieving. Intriguingly, no matter where in the world you go this always tends to have similar sonic properties, being both low-pitched and in the minor mode.

on health across the lifespan. What is more, it may be especially important for those in care. Over the years Tiffany Field, a Florida-based researcher and head of the Touch Research Institute at the Miller School of Medicine at the University of Miami, and her colleagues have published many studies demonstrating the therapeutic power of interpersonal touch.[58] Nevertheless, many in the scientific community have remained sceptical. Part of the problem has been that no underlying neurophysiologically plausible mechanism had been suggested to explain such results. But that has started to change. Hairy skin, that is basically all the skin on your body with the exception of the soles of your feet and the palms of your hands, is innervated by its own unique sensory system. The recently discovered C-tactile afferents that are found in the hairy skin respond preferentially to slow, gentle stroking, at the speed of a normal caress (about 3–10cm/1–4in per second). Not only is the stroking, or caressing, of the skin subjectively pleasurable, it can also result in the release of oxytocin and μ-opioids.[59]

Interpersonal touch can help us relax and sleep better, alleviate pain and fight off infection. At the same time, there are now far greater concerns about inappropriate social touching that tend to limit its provision, except under carefully prescribed conditions. But touching can also take a toll on those who have to provide it on a regular basis, such as nurses who are required to massage patients therapeutically. Delivering interpersonal touch to someone we do not know well is not an emotionally neutral activity. One solution might be to automate the process, perhaps robotically, or use a massage chair, although the research shows that artificial stimulation doesn't seem to have quite the same effect as interpersonal touch. As yet, researchers are not sure what is missing. It may be that in order to be effective, touch also needs to be warm like human skin (robots, remember, tend to be cold-blooded), or because the genuine emotional concern, interest or empathy signified by stroking is missing. And it should, of course, never be forgotten that anyone who is close enough to touch is also close enough to smell. It may well turn out to be that the combined influence of olfactory or pheromonal cues and gentle, warm tactile stimulation synergistically delivers the biggest benefits for well-being.[60]

In the nursing home where my mother ended her days, the care staff would periodically arrange for owls to be brought in for the residents to stroke. Although it was hard to tell, as Alzheimer's tightened its unrelenting grip on her mental faculties, my sense was that Norah enjoyed this periodic tactile interaction with nature. In fact, animal therapy is an increasingly common activity nowadays in a growing number of forward-thinking hospitals and care facilities. Being able to stroke and interact with another living creature really can provide some much-needed psychological comfort.[61] Nevertheless, given the emerging body of scientifically credible research documenting the beneficial effects of stroking so-called hairy skin (regardless of whether it is actually hairy), it is hard not to want to give that long-neglected area the attention it so obviously deserves. Stimulating the skin, or rather the C-tactile afferents, in other words, should perhaps be considered a biological necessity, and not merely a luxury for those wanting to be pampered.

## Scent-sory healing

Remarkably, even something as simple as releasing a sweet smell (think of something like caramel or vanilla) has been shown to help people to cope with pain. Thus, the sense of smell can also play a surprisingly important, if, once again, rarely acknowledged, role in healthcare. This ranges from the beneficial effects of aromatherapy massage through to masking unpleasant odours.

Back in the 1960s, one commentator was already encouraging hospitals to experiment with 'odour therapy' by blowing pleasant scents into the wards so as to promote feelings of security and well-being amongst patients.[62] In one Australian study, undergraduates were able to withstand the pain of an ice-bath, what is known in the business as the 'cold pressor test', for significantly longer with the smell of caramel in their nostrils than when either no scent, an unpleasant smell (civet), or a pleasant but not-sweet scent (aftershave) was presented.[63] This research builds on the observation that neonates will put up with the painful heel-prick procedure with less fuss if they have been given sugar. Sweetness, in other words, no matter whether tasted or just smelled, appears to have analgesic properties, resulting in reduced crying in newborns and increased pain tolerance in adults.[64]

Beyond the example of the ice-bath, there are now many studies demonstrating that scent can be used to help relax us and reduce stress.[65] For example, anxiety levels amongst women attending a dentist's surgery were found to be reduced when the typical eugenol/clove smell associated with fillings was replaced by an orange scent.[66] Note here, however, that it is not that the specific scent of eugenol is itself inherently, or innately, stressful. Rather this scent takes on a negative/stressful association as a result of our prior experience in dentists' surgeries.* That is, we soon learn that this scent tends to be associated with unpleasant sensations, hence making us anxious about whatever may be in store.† Presumably, though, after repeated visits to an orange-scented dentist's surgery this scent too would become negatively

---

\* Remember how in the Commuting chapter we saw that new-car smell is liked so much precisely because of its association, in this case with a high-value purchase.
† Intriguingly, females tend to be affected more by ambient scent than males in many of these studies, perhaps hinting at an enhanced sensitivity to olfactory stimuli.

associated, so perhaps the best solution would be to change the scent in the surgery every six months or so, which is the usual interval between check-ups.*

## *Multisensory medicine: processing fluency and the dangers of sensory overload*

While the research on sensehacking healthcare tends to proceed one sense at a time, in any realistic environment, multiple sensory signals will usually be competing for our attention simultaneously. As such, the effects of one sense hack cannot be properly considered without also thinking about what is going on in the other senses. And when multiple sensory cues are combined, there is always a danger of sensory overload. This may help to explain the results in a study conducted in a plastic surgeon's office in Germany. While the addition of a lavender scent or instrumental music with nature sounds reduced anxiety for those waiting for their appointment, combining these sensory cues eliminated the benefits.[67]

The idea of restorative, or calming, multisensory environments has been most extensively developed by those working with the Snoezelen concept. This term, which is derived from the Dutch verbs 'to explore' and 'to relax', refers to an approach that focuses on the development of controlled multisensory environments that are meant to be both relaxing and, at the same time, stimulating. They incorporate colourful patterns and lights, scents and

---

* One downside of this approach is that it doesn't offer much scope for the introduction of a signature scent of healthcare. Given the increasing use of such scents in so many other commercial settings, from hotels to shops, it can't be long before someone comes up with the distinctive branded scent of some premium healthcare provider or another. I wonder what Medicaid in the US or the UK's NHS would smell like.

music, as well as various materials to touch.⁶⁸ Originally, the idea was to provide some kind of intriguing and yet relaxing stimulation for people with special needs, such as those suffering from particularly severe brain damage. However, the approach has since been extended to other groups. The available research provides some limited support for the suggestion that such unstructured multisensory environments can exert a beneficial effect on the behaviour of various patient groups, including post-partum mothers, those suffering from dementia and certain groups of psychiatric patients.⁶⁹

In a way, then, this brings us back to the nineteenth-century notion of chromatherapy. For while light and colour undoubtedly do move us, it is likely to be the total multisensory environment, involving as it does sight, sound, touch, scent and perhaps even taste, that will have the biggest effect.

And remember, there is nothing to stop us all from picking our own playlist of therapeutic tunes next time we go into hospital for some planned treatment, taking a relaxing scent with us to the dentist's surgery, or giving our loved ones in care something to touch, or better still, touch them ourselves. It is the very least we can do to help sensehack our health and well-being.

# 9. Exercise and Sport

What actually determines how successful, or energetic, our workout is? Does it depend solely on our intrinsic motivation or is there anything else that we can do to help increase its effectiveness? What if something as simple as wearing red, or perhaps smelling some peppermint, could do the trick? As we will see, it is not only the muscles, heart or lungs that fundamentally limit our athletic performance, but also our brain. While our internal drive certainly plays its part, the multisensory environment in which we exercise has more of an influence on us than most people realize. By optimizing the exercising environment, be it outdoors in nature or else in one of the nightclub-themed gyms that seem to be popping up all over the place these days, any one of us can help 'nudge' our own physical and mental well-being in the right direction.

By hacking the senses, in other words, we can optimize the environmental stimulation and, by so doing, get more out of every exercise session. You may also find that you end up exercising more frequently. And while some of the sense hacks are intuitively obvious, such as listening to loud, fast, motivational music, a number of the others are not. And were you wondering why professional footballers periodically come over to the touchline, take a mouthful of some sports drink or other and then spit it out? Surely you have to swallow the drink in order to get the benefits, be it in terms of hydration or an energy hit? The fact of the matter in this case turns out to be stranger than fiction: the science shows that it is actually sometimes better to spit than to swallow. The footballers' senses are being hacked to help improve their performance based on rigorous science. But it is not only the professionals who can benefit.

There are a number of good reasons why we should all be exercising more than is currently the case: everything from doing our bit to help fight the growing global obesity crisis through to improving our cognitive performance by means of aerobic exercise.[1] Exercise provides an effective means of dealing with many of our modern maladies. It can, for instance, help restore our mental resources and, at least according to the results of a recent Cochrane Review, it may even help reduce the likelihood that we will suffer from depression. Meanwhile, a recently published study led by former Oxford psychology student Adam Chekroud found a 43 per cent reduction in self-reported days with poor mental health in those who exercised as compared with those who did not in more than 1.2 million North Americans controlled for age, race, gender, household income and education level.[2] That said, and despite the widespread and overwhelming support for its beneficial effects on both our physical and mental well-being, the evidence suggests that, on average, most of us simply do not get anything like enough exercise.

According to the 2008 Health Survey for England, only 40 per cent of men and 28 per cent of women meet the national recommendations of at least thirty minutes of moderately intense physical activity five days a week. In the United States, the figures are even worse, with only one in every five adults currently meeting the recommended guidelines. And when it comes to the elderly, only 17 per cent of men and just 13 per cent of women in the UK between sixty-four and seventy-five years of age meet the recommendations.[3] By the time the latest 2018 Health Survey for England came out, '27% of adults reported less than thirty minutes of moderate or vigorous physical activity per week and were classified as "inactive".' What is more, the report goes on to say, 'More than half of adults (56%) were at increased, high or very high risk of chronic disease due to their waist circumference and BMI.'[4] Nudging us all towards a healthier lifestyle could not be more important, and sensehacking offers one of the best hopes for addressing this mammoth challenge.

## Is it better to exercise in nature or indoors?

Given everything that we have seen concerning the nature effect, the answer to this question would seem obvious. Surely, we should all be exercising in nature wherever and whenever possible, not in some cavernous, dark, dank and sweaty gym. That, though, is often much easier said than done. After all, the 75 per cent of Europeans who now live in an urban environment may simply not have the opportunity to exercise out in nature on a regular basis. So, if the only practical choice is between exercising outdoors in an inner-city environment and working out in the gym, perhaps the latter really does offer the best alternative. The public would already seem to have made up their minds, with gym brands such as PureGym and Energie Fitness becoming increasingly popular. New facilities have been popping up all the time in most city centres. According to a report in the *Guardian* newspaper, growth in this sector has been exponential in recent years, with one in every seven people in the UK a gym member.[5] The latest reports highlight a similar trend in many North American cities.[6]

Enter one of these new-style gyms for the first time, though, and you'll be struck by the atmosphere. It really is a bit special, and not at all like the traditional brightly lit gym of old. Many look and sound a lot more like a nightclub, or an Abercrombie & Fitch store, than anything else, with loud, thumping, high-paced, motivational music and mood lighting.[7] These gyms, or at least the best of them, promise their members a carefully controlled multisensory environment that can, or so the claim goes, help those who join up to better reach their exercise targets. But what, exactly, does the science say about the impact of the environment on our motivation to exercise, not to mention the quality of our workout? And what kind of multisensory stimulation is best? To a certain extent, it depends on quite what you want to get out of your fitness regime. Let's kick off, though, by looking at the differences between exercising indoors and outside.

According to the research, running outdoors would seem to be preferred over doing the same indoors on a treadmill, though running in a park is significantly better for our mental (specifically emotional) restoration than jogging in an urban space.[8] Enriched natural environments can help to distract us from the sometimes unpleasant sensations that may be associated with physical exertion, and by so doing presumably support beneficial psychological outcomes. In one systematic review of nine separate experiments conducted in this area, moderately good evidence was obtained to support the conclusion that exercising in nature is more beneficial for us than running indoors.[9] Academics, you will understand by now, are not necessarily given to overstatement.

While exercising in nature may well be better for us in terms of restoring our mental capacities, if it is maximizing the intensity or duration of the workout that you are after, then the gym may be more conducive to helping you meet your goals, at least in the short term. In part, this is because the gym offers a more controlled sensory environment. Ultimately, though, regardless of what the multisensory atmosphere does to or for us while we are working out, what is perhaps more important than anything else is that we stick with the programme, whatever the fitness regime we have chosen for ourselves. On this front, the stats are pretty depressing, with almost half of us cancelling our gym membership within a year of signing up.[10] Potentially relevant in this regard is anecdotal evidence suggesting that we may just be that bit more likely to stick to our self-imposed regime in the long run if nature is involved.[11]

## Distraction for action

Many of us watch the TV while exercising in order to distract ourselves from what may otherwise be a boring activity. One bizarre attempt to do just this while out in nature came from the David Lloyd chain of gyms in the UK. In 2018, the company came out with the bright idea of sending its members out for a run behind a personal trainer carrying a small flat-screen TV strapped to their back. The idea was that the member, wearing wireless headphones, could catch up on their favourite TV show while being exposed to nature (I am being serious). You do have to wonder whether the benefits of 'the nature effect' still apply under such peculiar viewing conditions. And, if not, I suppose you might as well avoid all the hassle and simply stay indoors to exercise. Roland Gibbard, vice-president of the British Association of Road Racing, was even more blunt, describing this as a 'pointless exercise', going on to say, 'I think it is ridiculous and completely defeats the point of running outside. You may as well just stand on a treadmill.'[12] I couldn't have put it better myself.

When I first heard this story, I must admit that I had to check that it wasn't a hoax. This view was only reinforced a few days later when I came across another press piece titled 'The latest fitness trend – the cavewoman workout',[13] suggesting that an all-new 'ancestral health' movement involving 'exercise routines based on what humans naturally did 10,000 years ago, or moving like an animal' was the next big thing. Hmmmm, I have to say that I'm not too sure about that one either. It would seem to me that some people may be taking the ideas around evolutionary psychology and 'the nature effect' just a little bit too far. In truth, the David Lloyd story has more the whiff of a media-led marketing campaign than any serious attempt to change the way in which we exercise.

Rather than bringing technology or entertainment out into nature, others have been trying to bring nature, or a virtual version of it, indoors. For example, Art Kramer, from the University of Illinois, had people view a couple of large-screen displays showing nature scenes while they ran on an indoor treadmill. Others, meanwhile, have used VR headsets to provide an even more immersive 'outdoor' environment.[14] The movement of the scenery in these digital images, be it on the screen or in the headset, is then linked to that of the treadmill.* Sounds intriguing, right? But the question remains, can the benefits of exercising in a controlled indoor environment actually be combined with the nature effect? Unfortunately, early versions of these high-tech solutions have been beset by teething problems, as captured in Florence Williams's 2017 book *The nature fix*. When the writer tried out the latest technology needed to run the nature simulation in Kramer's lab, it was apparently both noisy, with a loud whirring sound in the background, and also subject to abrupt monitor resets.[15] Both of these factors are likely to spoil any sense of immersion in the virtual environment that the indoor exerciser might potentially have had. Should such problems be representative of the state-of-the-art then it could be argued that we have

---

* I can imagine that there is a very real danger here of runners falling off the treadmill too if they aren't careful.

## Mood music: moving to the beat

yet to see the full potential of virtual nature in the context of exercise.

In many sports, there is a well-established link between an athlete's mood, their level of anxiety and how well they perform. According to the results of one study, 45 per cent of the variance in the performance of elite male distance runners could be accounted for by their mood and anxiety.[16] While this has been explored most extensively in the world of elite athletics and professional sport, it would seem likely that a similar relationship holds true for the rest of us, no matter how gentle or vigorous our own workout.[17] What this means, in practice, is that any sensory intervention that helps us to relax, or else improves our mood, is likely to end up improving our (sporting) achievements too. We have already seen in the earlier chapters how various sensory interventions, including everything from the strategic use of music through to the release of ambient scent, influence both our mood and our level of arousal. One of the other things that sensory stimulation can be used for is to help distract us from the boredom, tiredness and/or pain that we may be feeling while exercising.[18] First, though, I want to take a look at how music can be harnessed to help us get the most out of exercise.

Music is probably the single most important sensory cue as far as sensehacking exercise is concerned. It motivates us and it may even be used to entrain (or synchronize) our own behaviour to the musical beat. No surprises for guessing that loud, fast music works best. Music provides a highly effective means of modulating our mood and emotions, and this, in turn, can influence our physical activity. Indeed, music, especially if synchronized with our own actions, leads to the release of hormones that can reduce the perception of strain and enhance the experience of positive emotions. Many published studies now show that listening to such music while exercising can improve performance and reduce perceived effort.[19] The effects here are not small either. Just take the results of one study demonstrating that runners listening to 'Happy' by Pharrell Williams reported enjoying their workout 28 per cent more than those who exercised in silence.[20] So, no matter whether you happen to be exercising indoors or out, why not think about selecting the music to help keep yourself maximally motivated? Just remember, though, that if you happen to find yourself somewhere scenic then the music might just end up *reducing* any benefits attributable to the nature effect.

Music not only motivates us while we are engaged in sport, it can also be used to psych us up before that big event. I can certainly still vividly remember my time as an undergraduate in Oxford, three decades ago now, when I used to be a regular down on the river. Before the big Torpids and Summer Eights rowing races, my crew would make a point of huddling together as a form of team-building pre-race ritual. Our coach, the improbably muscly North American Dov Seidman, would always have something like 'Eye of the Tiger' from the movie *Rocky III* blaring out. And while it might sound ridiculous, don't forget that the most decorated Olympian of all time, swimmer Michael Phelps, would always listen to a brash and aggressive hip-hop playlist poolside. If it worked for him then why not for the rest of us?

I am the first to admit that it is hard to imagine anyone getting worked up while listening to a Kenny G ballad, or the sounds of the dawn chorus. Silence wouldn't seem to be fit for purpose either. But that, of course, doesn't necessarily mean that louder equals better. After all, the evidence suggests that many of us are being exposed to noise levels that are actually damaging our hearing.[21] Someone really ought to tell all those gym instructors who are so fond of blasting out the tunes in their spinning classes. And no, I am not just being an old fuddy-duddy on this.

Just take the following results for evidence that louder isn't always better. Kreutz and colleagues reported that perceived effort and actual performance while cycling on an ergometer were not influenced in either high- or low-trained males by increasing the loudness of the electronic dance music playing in the background from 65 up to 85dB.* By contrast, increasing the musical tempo by 10 per cent resulted in people cycling harder and faster and enjoying the experience significantly more than when the tempo of the music was reduced by 10 per cent.[22] So, fast tempo but not too loud is what you are looking for if you want to hack your senses musically in the gym, especially when engaged in endurance events, and to a somewhat lesser extent high-intensity sports.[23]

Looking now specifically at exercising on a treadmill, the results once again reveal that listening to loud, fast music enhances performance.[24] For rhythmical activities like running, listening to music that can be synchronized with our own actions turns out to be most helpful.[25] According to research by Fritz and colleagues, 'musical agency' is key. That is, if people believe that their actions cause the beat in the music that they are listening to then they will perform better. These researchers equipped fitness machines with sound-processing software so that a person's movement during exercise controlled the production of synthesized sound, providing a sort of musical feedback in response to their actions. However, as soon as the participants

---

* It is worth noting that the 85dB loudness level falls well below the 100dB+ that one finds in many spinning classes. However, ethical constraints mean that the responsible research scientist is not allowed to subject their participants to such loud music, given that it may damage their hearing. All you spinning junkies out there please take note.

in this study were led to believe that someone else was responsible for the beat then the benefits of listening to the music dropped off. The researchers suggest that the role of agency in music creation facilitating behaviour may help to explain the emergence of singing and music sometime in the distant past. Presumably it may also help to explain the synchronized chanting that was once such a distinctive feature of prisoners working on chain gangs in North America too.[26]

We each have different preferences for everything from the tempo and loudness through to the style or type of music itself.[27] So, given its demonstrable impact on our performance, not to mention our differing musical preferences, it can't be too long before one of the music streaming services starts to offer personalized playlists designed to help you, or even perhaps to guarantee that you'll sweat more and/or burn more calories. In fact, Spotify has already started to move into this space. This seems like an absolute no-brainer to me. Sensehacking to make exercise easier and more enjoyable, now tell me who wouldn't want that?

There is also an emerging literature on the topic of sonification, referring to the provision of real-time auditory feedback (e.g. to athletes). Providing sonic feedback is sometimes more effective at improving people's performance than verbal or other types of feedback such as colour cues.[28] All in all, it can be argued that audition provides probably the single most effective means by which to hack our senses in order to improve our sporting performance. What is more, this is one of the situations where carefully chosen musical sounds, rather than the sounds of nature, turn out to be more effective. And, as we have just seen, listening to music influences us in multiple ways, from distracting us to entraining our behaviour to the musical beat, and from the benefits of sonification to enhancing our mood.

In many competitive sports, though, the only sounds that one usually hears come from the players themselves, or else from the spectators. Do such noises also affect sporting performance and, more importantly, can they be manipulated to deliver a competitive edge?

## Why do tennis players grunt?

Have you ever wondered why tennis players grunt on court? Maria Sharapova, the Williams sisters, Rafael Nadal and Novak Djokovic, to name but a few, are all (in)famous for doing so.[29] Prior to her fall from grace following a failed drugs test, Sharapova would scream at more than 100dB, leading fellow tennis player Greg Rusedski to suggest that she was 'louder than a 747' jet (though I suppose it all depends on how close you are standing to said plane).[30] Such noises are not simply the result of the player's physical exertion on court but may actually be serving a strategic role, by making it harder for their opponent to hear the grunter's shots.

In research conducted together with my colleagues at the University of Jena in Germany, we demonstrated – using a tennis game on TV that was stopped suddenly mid-volley – that where viewers believed the ball would end up depended, in part, on what they heard. When the loudness of the noise made by the racquet contacting the ball was amplified, people were convinced that the ball would bounce further into the opponent's side of the court than if the contact sound was made a little quieter. Note that participants could clearly see the ball being struck, and all they were asked to do was to indicate on a drawing of the court where they thought the ball would land. The sound of the ball's contact with the racquet, in other words, was technically irrelevant to their task. And yet, as we have seen time and again throughout this book, our brains can't help but integrate what is seen with what is heard, especially if the two sensory inputs seem to belong together. In this case, the cues were integrated in order to arrive at a judgement concerning the ball's trajectory.[31] That multisensory judgement, based as it is on information from both eye and ear, will usually be more accurate than a judgement call that depends on just sight *or* sound. In most situations, two senses really are better than one. It is just that in our study we deliberately distorted the sound in order to introduce a conflict between eye and ear. This is a favourite technique of research scientists who study the senses and their interaction.

While tennis players can't, as yet, change the sound that their racquet makes from one stroke to the next, what they can do is grunt loudly just as they strike the ball. Time it right and this noise will interfere with an opponent's ability to hear the contact sound, impairing their judgement of where the shot is going to end up and giving the grunter an unfair advantage.[32] Subsequent research from my colleagues in Jena has discovered that the grunt selectively impairs judgement of the length of the shot but not its angle, suggesting an integration-based rather than distraction-based explanation.[33]

No wonder that some commentators mutter about gamesmanship when discussing the grunters and screamers. Former World number 1 Martina Navratilova put it more bluntly still, asserting that grunting is 'cheating and it's got to stop'.[34] Noise is, in other words, more important than most people realize. Andre Agassi, one of the all-time tennis greats, recognized this after playing the inaugural match under the new roof on Wimbledon's Centre Court back in 2009: 'This was amazing. The way the ball sounds in here is going to add so much intensity for the players.'[35]

It is not just tennis, though, where sound matters. Expert basketball players second-guess their opponent's intentions better when they can hear their on-court movements,[36] and the golfers amongst you won't need reminding of the sweet ringing sound that a metal club makes following a perfectly struck tee shot. On hearing that noise, you don't even need to see where the ball ended up in order to know that it will be in a good place (bunkers permitting). However, since golfers do not compete in quite the same way as tennis players, there is no advantage in masking that contact sound from those they are playing with. Is it merely coincidence, do you think, that you have never heard a professional golfer grunt?

## Listen to the sound of the crowd

It is not just the contestants who are noisy. Very often, there is also the roar of the crowd to contend with. People often talk about home advantage, but what exactly has the sound of the crowd got to do with it? Intriguingly, the calls made by soccer referees are influenced by how much noise the crowd makes. The louder they are, the more likely a referee is to caution an offending player.* This, then, provides one explanation for home advantage in team sports, since the home crowd will protest more loudly than the away fans when one of their players is fouled due to sheer weight of numbers. The roar of the crowd really can influence what happens on the field.[37]

Interestingly, crowd noise affects referees' and judges' decisions more than it does the home side's performance. In one study, German football referees 'awarded' more yellow cards when viewing

---

* When team sports resumed behind closed doors during the Covid-19 pandemic, artificial crowd noise was pumped into TV and radio match broadcasts. This proved to be a great sense hack, at least for those who liked the idea. It would be interesting to learn whether home advantage and referees' apparent home bias disappear in empty stadia. It is interesting to note that crowd noise has been broadcast in empty stadia during cricket matches for the players' benefit.

video clips from games with the sound turned up high than when it was much quieter. Meanwhile, according to an analysis of all the European championship boxing bouts held over the last hundred years, 57 per cent of knockout blows between evenly matched boxers were delivered by the home fighter. (A knockout, note, provides a reasonably objective measure of the relative abilities of the two fighters.) By contrast, when bouts were decided by judges, the probability of a home win increased to 66 per cent for technical knockouts, and to 71 per cent for points-based decisions.[38] In boxing, then, the home advantage has at least as much influence on the judges and referees as it does on the boxers themselves. In fact, the home advantage tends to be more apparent in sports where judges or referees make the final decision, such as gymnastics and figure skating, than when objective performance criteria determine the outcome, as in weightlifting and short-track speed skating.

## *The scent of victory, the taste of success*

Not only can pleasant scents help mask the smell of all those other sweaty bodies in the gym, but releasing the right essential oil can also improve performance. In one study conducted in the US, forty athletes ran 400 metres on average 2.25 per cent faster with a peppermint oil-infused adhesive strip under their nostrils than without one. By contrast, the same aroma had no effect on the accuracy of basketball free-throw shooting, the latter, note, being a task based on skill rather than strength or endurance.[39] After a hard workout, we may well be sore, our muscles may ache and our joints may be stiff. So can the senses also be hacked to aid our recovery from exercise? Just think of it as a kind of 'sensual healing'.[40] As yet, despite some encouraging preliminary findings regarding the beneficial effects of aromatherapy, I would suggest that there isn't enough evidence to provide strong support for such claims.

Another of my favourite findings comes from sports psychologist Neil Brick and colleagues, who reported that club-level distance runners instructed to smile when exercising were able to make their physical exertion measurably more efficient, boosting the estimated economy of their running by more than 2 per cent. Such results might help to explain why the world's fastest marathon runner, Eliud Kipchoge, who in October 2019 became the first man to cover the distance in under two hours, always seems to be smiling when he races.[41] One can only wonder what top soccer coach José Mourinho's famously dour expression must be doing to his players' performance.

There is a whole industry promoting nutritional and protein supplements to build muscle and speed recovery after working out. However, what really caught my attention when thinking about the role that taste, or flavour, plays in enhancing our sporting performance was the effect on elite cyclists of gargling an energy drink.[42] It had already been shown that high-energy carbohydrate drinks improved cyclists' performance, a result that should surprise no one, given that energy drinks are known to help endurance athletes to replace glycogen, a form of glucose that is stored in the body to release energy.[43] Remarkably, however, Professor David Jones and colleagues subsequently discovered that if a cyclist just swilled a glucose or maltodextrin carbohydrate drink around in their mouth once every 7–8 minutes or so before spitting it out (so avoiding getting too full), their performance on a sixty-minute time trial also improved significantly. Simply *tasting* carbohydrate for a few seconds boosted the cyclists' exercise performance by 2–3 per cent. Perhaps this is why players in team sports are often seen doing this during breaks in play too.

But how could the cyclists' performance improve if they didn't actually swallow anything? Might it be that it is not the energy drink per se that enhances performance? (Note here also that simply injecting glucose directly into the bloodstream has no beneficial effect.)[44] One possibility is a phenomenon known as predictive coding – our brains predict the energy hit to come on detecting carbohydrate in the mouth, and it becomes a sort of self-fulfilling prophecy. Endurance-trained athletes optimize their body's performance based on the energy that they expect will soon be made available via their stomach. Intriguingly, neuroimaging studies have revealed that the brain areas involved in reward and motor control, including the insula/frontal operculum, orbitofrontal cortex and striatum, all light up in response to the carbohydrate taste. The suggestion is that this then makes exercise feel a little more pleasurable, or, at the very least, a little bit easier.[45]

As to whether our guts and/or brains would eventually get wise to this ruse is a question for future research. Similarly, we do not currently know whether less elite participants would benefit in the same way. Nonetheless, given that the benefit of carbohydrate rinsing has now been replicated in a number of studies, such findings represent a promising start for anyone wanting to hack their taste buds. Taken together, such results support what is known as the 'central governor hypothesis', which proposes that it is not the muscles, heart or lungs that fundamentally limit a person's sporting performance, but their brain. If true, there is all the more reason to think that the senses can be hacked to enhance performance.[46]

When thinking about hacking taste we should probably also mention gum. Many people chew the stuff believing that it helps them to control their stress, although research suggests it's the flavour-active compounds in the gum, rather than the repetitive mastication, that do the work.[47] So, perhaps without realizing it, we are back to the benefits of a minty aroma once again, mint being one of the most popular gum flavours. But beyond what you put in your mouth, it's what you wear that has been proven to have some of the most surprising effects on our sporting performance.

## The power of clothing

The Nobel prize-winning author Isaac Bashevis Singer was certainly on to something when once he wrote, 'What a strange power there is in clothing.' The clothing we wear while exercising really does make a material difference (sorry!) to how we perform, but perhaps not for the reasons you think. Rarely do we consider the clothes we are wearing, in part because the feeling against our skin normally fades from awareness pretty much as soon as we put them on. In fact, you probably weren't even thinking about the feel of your own clothes until I brought the subject up just now. But just because we don't pay much attention to this background tactile stimulation doesn't mean that what we wear has no influence on us. Sportswear brands like Nike stress the performance of their clothing, suggesting that wearing the right attire will help their customers better achieve their sporting ambitions. However, beyond the obvious ways in which the functionality of the clothing affects our athletic prowess, there are also a number of more surprising psychological effects to consider.

Anecdotal evidence hinting at the importance of clothing colour comes from Wayne Rooney, the former wunderkind of British football. The Manchester United and England striker would get upset if he didn't know what colour his team was going to play in the next day. Wayne's problem was that, for away games, Manchester United would sometimes play in red and sometimes in blue. Rooney would spend the evening before a big game imagining himself scoring the perfect goal (well, in fact, several of them), but in order for the striker to picture his own performance on the field he needed to know which strip he would be wearing:

> Part of my preparation is I go and ask the kit man what colour we're wearing – if it's red top, white shorts, white socks or black socks. Then I lie in bed the night before the game and visualize myself scoring goals or doing well. You're trying to put yourself in that moment and trying to prepare yourself, to have a 'memory' before the game. I don't know if you'd call it visualizing or dreaming, but I've always done it, my whole life . . . The more you do it, the more it works. You need to know where everyone is on the pitch. You need to see everything.[48]

Surely something as insignificant as the colour of the kit really shouldn't influence how many goals anyone scores, especially not a prolific international striker like Rooney, but maybe the logic works the other way round: perhaps England's top scorer was so prolific precisely because of all the sensory preparation – sensehacking, if you will – that he put into preparing for every game.

To some, I am sure, Rooney's approach to visualization might seem like nothing more than superstitious pre-match ritual. However, there is mounting evidence that recruiting sensory and motor imagery can, at least for those who have reached local club level, deliver more of a benefit than simply spending more time practising. After all, the expert has probably done enough of the latter already. Research shows, for instance, that basketball players who imagine making successful shots on court tend to perform better afterwards, scoring more free throws than those who spend an equivalent amount of time practising out on the court.[49]

Beyond its role in visualization, the specific colour of their apparel can also affect competitors' performance in many contact sports. For instance, professional ice-hockey and American-football teams wearing black tend to be more aggressive than teams wearing any other colour. What is more, teams that switch to wearing black see an immediate increase in the number of penalties awarded against them.[50] Combine this with the fact that adopting an expansive body posture also affects one's sense of power and action tendencies,[51] and you have got yourself an explanation for the pre-match haka ritual performed by New Zealand's All Blacks. So, if you've ever wondered how this small island nation's rugby team so frequently manages to rank the best in the world, now you know: sensehacking at its very best.

## Seeing red

In one oft-cited study, anthropology researchers from the University of Durham analysed the results of all competitive bouts from four men's combat sports at the 2004 Summer Olympic Games in Athens: boxing, tae kwon do, Greco-Roman wrestling and freestyle wrestling. In the Olympics, competitors in all these contests are arbitrarily allocated either blue or red bodywear. Those in red were slightly, but significantly, more likely to win than those in blue. Furthermore, and as one might have expected, the benefits were most apparent in those contests where the opponents were most evenly matched. If the skill difference was too great, the colour of the strip had no effect. Under such asymmetric conditions, raw talent unsurprisingly ruled the day.[52]

The same researchers went on to look at performance in another competitive sport: football. This time, they analysed all the games played in the Euro 2004 tournament in Portugal. The preliminary evidence suggested that the five teams who habitually wore red did slightly better when playing in red than when playing in shirts of any other colour. That said, the sample size in the latter analysis was pretty small. Subsequently, though, confirmatory evidence emerged from a systematic long-term analysis of English football league results stretching back to the 1946–7 season.[53] Teams with a red strip consistently outperformed those wearing another colour. In fact, the same result was observed in every division of the league. Red teams were also more likely to win the league title. When playing away, teams must sometimes play in a change strip. The researchers were able to use this fact to demonstrate that it was the colour of the clothing, not the inherent qualities of the team concerned, that made all the difference. That is, 'red' teams only performed better than expected in their home fixtures and not in away games (when not wearing their red strip).*

It is not just the players who are influenced by the colour of the kit; referees are affected too. In one study, researchers from the University of Münster used graphics software to reverse the apparent colour of the players' protective gear in a number of four-second video clips taken from tae kwon do sparring rounds. The forty-two experienced referees who took part awarded an average of 13 per cent more points to a player in red than when an identical clip showed the same player wearing blue.[54] Given that the alternative account for the red effect in terms of visibility differences has now been ruled out, a psychological/hormonal explanation for the most surprising, yet seemingly ubiquitous, red effect seems more likely.[55] Furthermore, this colour doesn't just affect the outcome of sporting contests. It has also been shown to impair people's performance across a range of other situations, including using a red pen to complete an IQ test as well as in other achievement contexts. This has led some researchers to the conclusion that 'seeing red' may trigger what is known as avoidance motivation.[56] The

---

* Though here, of course, one should perhaps not forget the 'sound of the home crowd' advantage that we came across earlier.

take-home message here is that subtle sensory cues alter the outcome of professional sporting encounters and, what is more, there is no reason to believe that such sensory influences should be restricted to elite athletes.

As always, there is a nice evolutionary story around these findings.* In nature, the presence and intensity of red coloration are thought to act as an evolutionarily important signal associated with dominance, arousal and aggression. As Hill and Barton note, 'Red coloration is a sexually selected, testosterone-dependent signal of male quality in a variety of animals.' Wearing this colour, then, may trick your, or your opponent's, brain into thinking you are a little more dominant, as dominant male animals look redder, on average, while submissive or scared creatures pale in comparison. Consistent with this view, no red advantage was observed amongst the female competitors in the contact sports from the 2004 Olympics.[57] The evolutionary importance of being able to 'read' these subtle variations in skin tone is further hinted at by the observation that our trichromatic visual system appears to be more sensitive to those hues associated with the flushing or blanching of the skin than to any other colour in the visible spectrum.[58]

The US military also conducted research into clothing. Hard though it is to believe today, there was once a belief in certain quarters that wearing red underwear would embolden any warriors whose tackle was so encased ... Except, of course, that it didn't! As I am sure you might have guessed, carefully controlled scientific research failed to provide any support whatsoever for this particular sense hack.[59] Given what we have seen in this section, it was presumably the fact that the enemy fighters couldn't see the colour of their opponents' underwear that explains why

---

* In fact, one sometimes wonders whether there is anything that evolutionary psychologists can't explain.

this particular red ruse didn't work. And so this, in case you were wondering, helps explain why Superman would always wear his red underpants on the outside for all the world to see. Perhaps not quite such a silly idea when you think about it!

However, I do have an amazing tip for those looking for a psychological boost coming from the latest tranche of strange but (probably) true scientific findings from the emerging field of 'enclothed cognition'.[60] This relates to the idea that what we wear influences how we think. Of course, it also affects how others respond to us. Extrapolating from the results of one popular science study, if you were to play sport wearing a Superman T-shirt, even if no one else knew that you had it on, then you might find that you performed better than if you were to wear something else. In the underpinning research, wearing superhero attire enhanced students' self-esteem and how much weight they thought they could lift. Extending the logic, one might therefore wonder whether wearing a Spiderman T-shirt would promote rock-climbing prowess. That said, it is important to bear in mind that 'the superman effect', while extensively covered in the press,[61] has, at least so far as I am aware, not yet appeared in a peer-reviewed journal. So, given the reproducibility crisis in science these days, and bearing in mind the explanation for the null effects of red underwear we came across a moment ago, you might just want to wait a while before trying to scale the nearest cliff face wearing your favourite superhero attire.

## *Working-out with the senses*

Given that most of the research in this chapter has involved hacking just one sense at a time, it would be really interesting, looking to the future, to combine the various sense hacks. Could the performance-enhancing properties of peppermint aroma be combined with up-tempo music? What will be needed in the years ahead are more studies where researchers systematically manipulate the input to several of the senses simultaneously so as to determine whether, for example, the performance or endurance boost is bigger. I certainly believe that the benefits would soon start to stack up.

While there is probably no need to invest in a pair of red undies just yet, it pays to choose one's colours carefully in a competitive physical sport like wrestling or soccer. It won't necessarily help you to punch above your weight, but it might just give you the competitive edge that you need against an evenly matched opponent. It is only by hacking the senses, after all, that one can really hope to maximize the benefits of exercise, be it for fitness, health and/or our mental well-being. Throw in a glucose mouthwash too, and, if you don't feel too embarrassed, a Superman outfit, and who knows how much better you might perform?

# 10. Dating

Beauty is big business these days. Who, after all, wouldn't like to be more attractive? However, as we will see in this chapter, attraction is much more than merely skin deep. It is, in a very real sense, a multisensory construct.[1] But which of your senses do you think is more important when it comes to conveying a sense of beauty? How exactly do people use their eyes, ears and nose when evaluating a potential mate, and do men and women rely on their senses in quite the same way? And what should you do if your senses give you conflicting messages? To give you an everyday example, how would you feel if you met someone who looked absolutely fabulous but whose smell you didn't like quite so much? These are just some of the intriguing questions that I want to address in this chapter.

According to the evolutionary psychologists, we are drawn to fitness – evolutionary fitness, that is. People who look healthy are more attractive to us because they represent a better biological prospect in terms of their reproductive potential. Now, there is not much that we can do about the genetic hand that fate has dealt us, other than perhaps to blame our parents. And there are no prizes for guessing that many of the sense hacks around boosting attractiveness work by accentuating, or caricaturing, those natural signals of evolutionary fitness – think lipstick, eyeliner, high heels, red clothing and push-up bras – or, in some cases, masking them – think perfume, deodorants and razors. But unlike plastic surgery and Botox, the suggestions that you will find in this chapter should keep on working, year after year.* Curiously, it turns out that we are often not aware of the sensory cues that most influence our judgements. First, though, before we get to all that, I want to tell you about one of the simplest hacks that has been shown to increase how attractive you appear.

## *Arousal*

One of the most effective ways in which to increase your attractiveness to others is by getting them aroused. No, not like that! The basic idea here is that people are not always very good at attributing the cause of their own arousal correctly. We tend to misattribute it to whomever we are interacting with at the time, rather than to the true environmental stimulus underlying our altered state. For example, in a classic study from the field of social psychology, published in 1974,[2] Canadian researchers had a young female interviewer approach several single young men while they were crossing a bridge and ask them to fill out a short questionnaire. Two bridges were used: one was a fear-inducing wobbly suspension bridge, the other much more substantial. When the forms had been completed, the interviewer tore a corner off and wrote her name and number on it, offering to explain more about the study at a later date. The researchers wanted to know whether the number of participants who would call afterwards would be affected by which bridge they had been crossing (the expectation being that the young men would be more interested in

---

* And if you thought that Botox was just for humans you'd be wrong. In 2018, twelve proud Saudi Arabian camel owners were disqualified from a competition after having given their star animals injections in order to improve their pouts (see www.theguardian.com/world/2018/jan/24/saudi-camel-beauty-contest-judges-get--hump-botox-cheats).

trying to date the interviewer than in the science underlying the study).* The results were clear: nine of the eighteen men on the scary bridge phoned, as compared with just two of the sixteen who had been approached on the other bridge.†

Other research has demonstrated similar arousal-induced attraction effects in individuals stepping off a rollercoaster ride when compared with those about to board.³ Watching an arousing movie increases affiliative behaviour amongst couples too.⁴ One might, I suppose, consider whether boy racers in their boom-box cars are using a similar strategy, albeit one that is presumably based on intuition rather than empirical evidence. After all, driving dangerously is likely to arouse whoever else is in the car, as is the loud music. They may be wasting their time, though, at least if the results of the original scary-bridge experiment are anything to go by, as the misattribution of arousal only worked on young men, not young women. Nevertheless, next time you find yourself on a flight sitting next to an attractive fellow passenger, my advice would be to wait until you hit some turbulence, the rougher the better in this case, before making your move. You might also want to take your date to a thriller movie next time you have ulterior motives for wanting to visit a cinema.

My colleague Helmut Leder from Vienna has demonstrated that we tend to misattribute the arousal induced by listening to music as well. Leder and his fellow researchers had people rate the attractiveness of a series of professionally photographed pictures of unfamiliar individuals of the opposite sex posing with a neutral expression, either in silence or while listening to

---

* This, like most of the other research that we will come across in this chapter, is focused on heterosexual attraction, I'm afraid, as this is where, historically, the vast majority of this research has been undertaken.

† If the pun weren't in such bad taste, you might even say that those in the 'scary bridge' condition had jumped at the chance.

nineteenth-century piano music, varying in terms of its pleasantness and arousal. Women tended to rate the male pictures higher after having listened to music as compared to when rating the faces in silence. What is more, listening to music that was highly arousing had the largest effect on ratings of both facial attractiveness and dating desirability.[5] In this study, though, the male participants' ratings were unaffected by the musical intervention.

Do such results, I can't help but wonder, help to explain some of the popularity of clubbing amongst young people? Does all that loud arousing music (remember the boy racers) make men look a little more attractive, hence facilitating social interaction? Quite possibly.[6] Synchronized gyrating also helps.[7] Evolutionary psychologists have long argued that dancing plays a role in mate selection,[8] but what exactly should those who want to polish up their dance steps do to improve their chances? The good news is that a number of serious scientific analyses now provide tips on exactly how to 'shake your stuff' in order to make your moves maximally attractive.[9] Bear in mind, though, that men and women tend to be looking for somewhat different things.

Men hoping to appeal to women should note that it is the variability and amplitude of your neck and trunk movements as well as, wait for it, the speed with which you move your right knee that matter most. (*What?* I hear you say. At least I don't feel so bad about never managing to get the hang of dancing.) The more varied and pronounced the movements, the better. Women apparently consider these to be signs of genetic quality that indicate health, vigour and strength. Meanwhile, when men rate the quality of female dancing, they tend to be impressed by a greater amount of hip swing (the Colombian singer Shakira clearly appreciates this), more asymmetric thigh movements and intermediate levels of asymmetric arm movements. So now you know what to do if you want to hack a potential mate's senses while you are on the dancefloor! Though, next time you get lucky, just remember that it may be the music as much as your coolest moves that is really giving you that little extra boost.

Researchers have also looked into sexual signalling in the context of the discotheque (or clubbing). According to one revealing analysis, ovulating Viennese women (i.e. those at the most fertile point in their cycle) going out without their partners were found to dress a little more provocatively.*[10] Women's faces also tend to be rated as looking a little more attractive during the most fertile phase of the menstrual cycle,[11] thus perhaps helping to explain why professional lap dancers make nearly twice as much in tips while ovulating than while menstruating.[12]

It is, however, not only swinging rope bridges and arousing music that work.† When I was younger, so much younger than today, I used to cook some very spicy meals for certain lucky ladies of my close acquaintance. Now, I must admit that I haven't had the chance to perform any statistical analysis on this, and I fear that experiment's sample size might have been a little on the low side.‡ Nevertheless, my feeling was always that the flushed skin, sweating, palpitations and dilated pupils that were such a common response in those who were subjected to one of my legendary spicy Thai green curries or fiery pasta arrabiatas would get my lady friends' brains confused. Perhaps those who were so afflicted would misattribute their most

---

* The findings would presumably also hold for women from elsewhere too. And should you be wondering how exactly the researchers went about assessing the provocativeness of the ladies' attire, they write: 'We digitally analyzed clothing choice to determine the amount of skin display, sheerness, and clothing tightness.'

† Mrs Spence, it's OK, you can skip this paragraph!

‡ On second thoughts, perhaps I shouldn't describe it as an 'experiment', otherwise I'll have my University Ethics Panel breathing down my neck demanding to see the ethical approval!

unusual bodily sensations, or should that be symptoms, to their charming and debonair chef for the evening, rather than to whatever was lurking on their plate. More often than not this gastrophysics-inspired seduction strategy seemed to work. So much so, in fact, that this was one of the top tips I passed on when consulting for Heston Blumenthal's Valentine's Day-inspired *Recipe for Romance* on TV.

## 'The look of love'

What do we find attractive in a potential partner? As I mentioned already, it hinges on evolutionary fitness. Sorry to be so unromantic about it, but that is the unavoidable conclusion of several decades of research in the field of evolutionary psychology. For both sexes, and regardless of one's sexual orientation, the more symmetrical your face, the more attractive you look.[13] Left/right symmetry is an important sign of evolutionary fitness. It is preferred not just because it is undoubtedly easier on the eye, but also because it signals a potentially healthy mate. Asymmetry of face, or form, by contrast, often indicates that whoever is being evaluated has suffered some unfortunate environmental impact(s) along the way. Their goods, as it were, appear damaged. Imbalance is linked to age, disease, infection and parasitic infestation, none of which is especially attractive in my book. If I could take your face and average the two sides, making you look a little more symmetrical in the process, you would be rated as more attractive by others; it is as simple as that.*

---

\* Sadly, though, you can't do the same thing to synthesize a more attractive voice. Try averaging the frequencies of different voices and what you will be left with is lots of unpleasant harmonics.

Smiling is taken as a sign of good health and hence is also deemed attractive.[14] In 2010 it was reported that the Major League Baseball players from the 1952 draft who smiled more intensely in their photographs from the time tended to outlive those whose grin was a little more subdued.[15] That said, the suggestion that happy people might actually live longer, all other things being equal, has subsequently been brought into question by the results of a huge study, reported in *The Lancet*, of more than 700,000 women across the UK, which uncovered no evidence that happiness per se increased a woman's longevity, when other factors such as general health and sleep quality were controlled for.[16]

One other tip here for those wanting to improve the impression they give is to make eye contact, be it in reality with another person or virtually while gazing into a camera. The reason being that this is both more arousing and more attention-capturing than a deviated stare. What is more, the neuroimaging research reveals that you will better activate a viewer's ventral striatum, one of the brain areas involved in reward prediction, by looking directly at them, even if only gazing out from a photograph or screen.[17]

Another way to gain a competitive advantage in the mating game is to play the guitar,\* or, at the very least, look like you do. In one small online study the likelihood that female students would respond to a Facebook friendship request from a young man increased when he was pictured holding said instrument.[18] On reflection, perhaps this result should not come as such a surprise, given that music-making is thought to signal evolutionary fitness, implying, as it does, a degree of both creativity and manual dexterity.[19] Charles Darwin certainly thought that

---

\* This sense hack should presumably work with any other instrument too.

The Venus of Willendorf.

the evolution of certain male traits might be explained by sexual selection via mate choice. Note that music-making fits right in here as a kind of courtship display that varies in terms of both its emotion and its complexity.[20] Intriguingly, the more complex the music, the better/more attractive the player, at least if you ask a woman who is ovulating.[21] At the same time, however, creativity can also make up for low physical attractiveness in both sexes.[22]

From the heterosexual male perspective, wide hips for childbirth and large breasts for feeding any offspring, should things get that far, have long been recognized as signs of universal fertility. Just take the amply equipped Venus of Willendorf, one of the earliest fertility symbols. There is certainly little danger of this Venus going hungry any time soon.

Going in another direction, though, it has been argued that women tend to be less interested in looks and rather prioritize the smell and sound of potential mates.[23] In this respect, at least, men really are from Mars and women from Venus. Once again, this makes perfect sense from the perspective of evolutionary psychology in that men can do a pretty good job of assessing the potential quality of a mate, which boils down to youthfulness and health, by eye alone (though, as we will see in a moment, olfactory cues turn out to be important too). By contrast, for women, a potential mate's natural scent is the single most important sensory cue to his immunological profile, informing her probably subconscious assessment of offspring viability. And, lest we forget the obvious, being rich appears not to harm a man's reproductive chances either.[24]

Men with a masculine face and/or voice tend to be attractive to women. As such, the ratio of the second to the fourth finger, what is known as the 2D:4D digit ratio, ought to be sexy too, especially when the ratio is low. This indicator provides a measure of a man's masculinity, as determined by testosterone exposure during early pregnancy. If a man's ring finger is longer than his index finger then his 2D:4D ratio falls below one. The average ratio amongst men is 0.98. The lower the ratio, the higher the level of foetal testosterone, and the more masculine the traits that an individual will exhibit.[25] The lower the ratio, the higher the sperm count per ejaculate too.[26] According to the research, long-distance runners and elite musicians both exhibit a lower ratio than average. However, while this might lead you to believe that heterosexual women should really be on the lookout for potential mates with a low 2D:4D ratio, it is important to note that there are also a number of potential downsides associated with high levels of prenatal testosterone, including an increased risk of autism, dyslexia, migraine and a compromised immune system.[27] Perhaps this trade-off explains why this sexually dimorphic sign (being more apparent in men than in women) is simply not particularly 'sexy'. Is it merely coincidence, though, that the fourth digit is the ring finger in the majority of cultures, announcing that the wearer is engaged, married or at least pretending to be?*

## Scent of a woman: sexy smells

A woman's natural smell changes subtly over the course of the menstrual cycle.[28] Men find a woman's underarm odour more

---

* The ancient Egyptians are thought to have been responsible for this habit, since they believed, mistakenly as it happens, that the fourth digit of the left hand contained the *vena amoris*, a vein that supposedly carried blood straight to the heart.

pleasant, more attractive and less intense at the follicular (i.e. fertile) phase of the cycle than when she is menstruating.[29] There is, in other words, more sensory information potentially being 'transmitted', or 'leaked', as the authors of some of the studies in this area describe it, than we realize.

Amazingly, we can even sniff out some useful information concerning a person's personality, namely their extraversion, neuroticism and dominance (three of the Big Five personality traits), based on nothing more than their body odour.[30] Beyond the attractiveness/pleasantness of a woman's odour, both men and women are able to smell the innate immune response that provides an early chemosensory cue of sickness.[31] It is also possible to infer something about a person's age from their bodily odour.[32] We can even sniff out the likely quality of another person's dance moves. In fact, those whose odour we like tend also to display bodily movements that we find attractive.[33]

Both men and women are sensitive to that part of another person's DNA known as the major histocompatibility complex. The suggestion is that we are not romantically attracted if someone's scent is too similar to our own. This helps us to avoid inbreeding by advantaging diverse mating partners.[34] That said, we generally find the smell of other races less attractive than our own, so such diversity does not always extend to ethnicity. Nor, in fact, to facial attractiveness, where we are actually drawn more toward those who look just like us. Given all the above, it might seem strange to realize how much time most of us spend trying to eliminate our own body odour with fragranced products, and that our attempts to do so have made the fragrance industry a multibillion-dollar business.[35]

Have you ever wondered why different people prefer different fragrances? Is this just random variation? Or could it be that the fragrance we choose in some sense amplifies our own natural scent profile? A few years ago, a speaker in an industry conference suggested that our choice of fragrance might actually be linked to our body odour. Should such a claim be even partly true then we might all be revealing a little more about ourselves than we realize. Intriguingly, Lenochová and colleagues have reported that the smell of a person's body odour mixed with their preferred perfume was rated as more pleasant by others than a blend of their own body odour mixed with a randomly allocated perfume, even though the perfumes themselves didn't differ in pleasantness when assessed individually.[36]

## 'The Lynx effect'

'The Lynx effect' has been the strapline for Unilever's bestselling deodorant brand for more than a quarter of a century.* So effective has the marketing campaign been that there was almost no desire from those in charge to want to investigate whether the claim was true. The success of the campaign, at least amongst the target group of young men, is amply illustrated by the complaints on various online forums from exasperated school teachers sick of smelling it in their teenage classrooms first thing in the morning.[37] You have to feel sorry, though, for the sixteen-year-old New Zealander Jayme Edmonds who used Lynx to polish the dash, and to remove paint stains from the door lining. He then went to his car in his boxers in the middle of the night to listen to the stereo while having a smoke. Bad idea, very bad idea. Lynx, like many other aerosols, is highly flammable. The car burst into flames, leading to three days in hospital followed by nine more days off school recovering from the blast. Not his lucky day, that's for sure.[38]

* The brand is known as Axe outside of the UK.

After a few years of pestering, I finally got the green light to put the Lynx effect to the empirical test, using a database of anonymous male faces that had been carefully selected to cover the full spectrum of male attractiveness – think John Belushi at one end and George Clooney at the other. We had a group of young ladies rate those pictures in terms of their attractiveness. Even though our participants were only ever asked to report on the eye-appeal of the faces that were flashed up briefly on the computer screen in front of them, their judgements were nevertheless biased by the ambient scent. Smelling Lynx deodorant resulted in the men's faces being rated as slightly, but significantly, more attractive than when either no odour or else an unpleasant odour (either synthetic BO or burnt rubber) was presented. As I am sure you can imagine, this reading of our results pleased our commercial sponsors no end. So far, so good. However, it turned out that a rose scent had about the same effect on male attractiveness. In other words, while the Lynx effect is real, the most parsimonious conclusion to emerge from our research was that a man can probably get exactly the same lift to his attractiveness by wearing any fragrance, just so long as it smells pleasant to whomever he is trying to impress.[39]

Gourmand fragrances, scents that smell good enough to eat, have become popular in recent years.[40] A few years ago Unilever jumped on this particular bandwagon with their chocolate-scented Lynx range. I wonder what this particular smell would have done to the ratings in our Lynx study, especially given the results of a Valentine's Day report appearing in *The Economist* highlighting the existence of a positive correlation between the amount of sex a nation gets and how much chocolate it consumes.[41] Of course, that said, it should be remembered that correlation isn't causation.

In a follow-up to our behavioural study, this time conducted in the brain scanner, we were able to demonstrate that the pleasant scent of Lynx actually changes neural activity in those parts of a woman's brain that code for the attractiveness of a man's face.[42] For this research, we first identified the particular region of the orbitofrontal cortex (OFC) that responds to the attractiveness of male faces. We then carefully analysed the brain's response to the full range of faces when presented in isolation. This revealed a focus of neural activation localized towards the middle of the OFC for more handsome faces. By contrast, those faces categorized as less attractive shifted the focus of activation towards the periphery of this brain structure. Crucially, the presence of a pleasant scent shifted the neural response towards the attractive portion of this reward-related brain area. However, perhaps the biggest impact of the research was when an article appeared in the bestselling lad's magazine *Maxim*, telling its roughly 9 million young male readers which part of a woman's brain they should really be targeting (that is the OFC, of course).[43] Talk about impactful research!

But while many young men appear to believe that the stronger the fragrance, the more effective it is, research conducted in California shows that the opposite is sometimes true. That is, ambient scent can sometimes actually be more effective when it is presented just below the threshold for awareness, when people are not even conscious of smelling anything.[44] What this means, in practice, is that even if you can't detect any smell in the air you can never be entirely sure that you aren't being influenced by what your nose is picking up subliminally.

It is not just physical attraction that is affected by scent. The people we see in pictures can be made to look more masculine or feminine, more sympathetic or more emotional with the right scent added. In Oxford, we even helped women to look a little younger. Working with Japanese fragrance house Takasago, we were able to reduce the perceived age of middle-aged women by about six months simply by using one of the company's proprietary 'youthful' fragrances. To date, most of the fragrance and attraction studies have used static pictures of unfamiliar faces. Now, while this might be what you typically see on an online dating site (and there is plenty of interest in scent-enabled dating apps), fragrance may well prove to be a little less important when evaluating a familiar dynamic (i.e. moving) face. This is undoubtedly one of the important questions for future research.

## Why is the lady always in red?

It is not just what you look like that matters, of course, but also what you wear. Unsurprisingly, clothing plays an important role in determining both how attractive we look and how we feel about ourselves. And, given men's greater reliance on vision, it is perhaps no surprise to find that it is women's attire that is most often the subject of serious scientific scrutiny. Indeed, Griskevicius and Kenrick found that 'Women across the world expend a great deal of time, energy, and money choosing clothes, accessories, and shades of make-up that enhance their attractiveness.'[45] According to a number of, as it so happens, male researchers, women are typically rated as more attractive and more sexually desirable when they wear red. Men also attribute more sexual intent to those women they see wearing red, though without necessarily being aware of the effect that this most provocative of colours is having.[46] This kind of sensehacking normally goes on outside awareness, at least according to Professor Andrew Elliot, who visited the Crossmodal Lab in Oxford on his sabbatical a few years ago. Wearing red has no influence whatsoever on women's rating of either men or women.[47]

Elsewhere, researchers have used the size of the waitress's tip as an admittedly crude proxy for female attractiveness. Once again those wearing red, a T-shirt in this case, tended to walk away with higher tips at the end of the evening than those wearing a black, white, blue, green or yellow T-shirt.[48] Further analysis confirmed that shirt colour influenced the tipping behaviour only of men. If the waitress puts on some make-up, and perhaps a flower in her hair, the size of the tip increases still further.[49] Wearing cosmetics leads to increased ratings of attractiveness too, at least when measured by restaurant tipping behaviour.[50] That said, the latest research suggests that the lift to a woman's attractiveness that can be achieved by the use of cosmetic products, even when the make-up has been applied professionally, while significant, nevertheless tends to be small compared with the differences in natural beauty between individuals. One of the experiments in this study tested a group of thirty-three YouTube models, while the other study involved forty-five supermodels.[51]

Intriguingly, Elliot and Niesta found that colouring the surround of a woman's photo red, as opposed to white, green, grey, blue or green, led men (but not women) to view them as more attractive and sexually desirable. Meanwhile, Taiwanese researchers have reported that men rate women carrying a red laptop as better looking, and as having more sex appeal, than when carrying a black, silver or blue laptop. In other words, it is not just red clothing or backgrounds that can be used to sensehack a man's judgements.[52]

Intuitively, people have known about the 'attractive red' phenomenon for millennia. After all, women have been using rouge and red lipstick at least since the time of the ancient Egyptians some 10,000 years ago. Furthermore, 'the lady in red' is a popular trope in cinema, song and even cereals. Think of Gene Wilder's romantic comedy *The Woman in Red* from 1984, or the red dress used to represent passion or sexuality in stage and film productions such as *Dial M for Murder*, *A Streetcar Named Desire* and *Jezebel*.[53] Chris de Burgh released his song 'The Lady in Red' back in 1986, and one often sees a red-frocked women pictured on the sides of boxes of Kellogg's Special K breakfast cereal. The character Jessica in *Who Framed Roger Rabbit?* even had red hair to complement her red dress.

However, despite its popularity, the 'attractive red' phenomenon is turning out to be one of a growing number of media-friendly effects reported by social psychologists that have come under stress. Indeed, at the time of writing, several research groups have tried and failed to replicate it, including amongst a large sample of more than 830 Dutch and North American men in one study. And wearing red doesn't always increase the size of the tips that waitresses collect either.[54] So, will wearing red attract a man? To be frank, it is hard to know quite what to say. The basic idea undoubtedly makes evolutionary sense, namely that red clothing simulates, in extreme form (i.e. caricatures), the red hue of the skin when we are aroused. That said, it is always important to remember that the meaning of red, or any other hue for that matter, is context-dependent, and this might be why red clothing doesn't always have the desired effect. It is also true that many of the significant results reported in the literature were obtained in studies with very small, so less reliable, sample sizes.[55] The jury is very much still out on this one, but let's just say that I don't think I'll be retiring my ubiquitous red trousers just yet. Long live sensehacking!

## Killer heels

High heels are, so I am told, the bane of many a woman's life. Some argue that they make women look more attractive by elevating their stature, making them appear more sophisticated. Given the influence that red might have on the male of the species, one can perhaps better understand why it is that the Christian Louboutin brand has been fighting quite so hard in recent years to protect the distinctive red coloration on the underside of their distinctive footwear. However, the latest evolutionary story here, backed up by some reasonably solid empirical evidence, is that what is also important is that high heels lead the wearer to arch their back a little, thus increasing lumbar curvature.[56] The optimal angle between the lower back and the buttocks is apparently somewhere around 45.5 degrees. This is supposedly so attractive to the male of the species because it mimics lordosis, the stereotypical posture seen in many species indicating sexual proceptivity.

At the same time, however, it may also reflect a morphological adaptation that provides a benefit to bipedal organisms during pregnancy. This is what is known as 'foetal loading'. Increased lumbar curvature might, in other words, be attractive to men precisely because it is associated with an increased chance of a mate sustaining multiple pregnancies without incurring damage to the spine while at the same time facilitating her ability to forage later into pregnancy.[57] Before lumbar curvature was accentuated by means of heels, evolution achieved much the same goal by wedging the third-from-last lumbar vertebra.[58] Whatever the cause, the novelist John Updike would seem to have perfectly captured the importance of this cue in his work *Pigeon feathers and other stories* when he wrote, 'A woman's beauty lies, not in any exaggeration of the specialized zones, nor in any general harmony that could be worked out by means of the sectio aurea or a similar aesthetic superstition; but in the arabesque of the spine.'

## Swipe right: tips for online dating

More dating probably takes place online than anywhere else. Sites and apps such as Tinder, Grindr and Ashley Madison cater to those in search of love (or lust) the world over. One market research poll suggested that seven in ten of those looking for a relationship would consider using online dating despite having never used it before.[59] Similar sites are used, or so I have been told, by those who are not so lonely but who simply want to hook up for coffee, conversation or something a little more – how shall I put it – entertaining. However, given that those who view your online profile will essentially only have one sense to go by, namely, what you look like in your profile pictures, then you obviously need to do everything in your power to make sure that you create the right visual impression in the second or so that anyone probably devotes to looking at you. After all, people make up their minds in as little as a tenth of a second. Gazing at a picture for any longer than that doesn't really change the visual impression you make, it merely increases other people's confidence in their judgement of you.[60]

But are you uploading the snaps that show you at your most attractive, assuming, say, that 'good-looking' is what you are trying to convey? According to the results of one recent Australian study, the answer is probably no. These researchers had more than 600 people upload pictures of themselves, and then rank their own snaps as well as the images that the other people in the study had brought in. Surprise, surprise, most young people did not actually pick the snap of themselves that others chose for them. So next time you decide to upload some pictures online, why not crowdsource your friends and ask them to pick the shot they prefer?* You might just find that you end up being that little bit more popular, or 'liked', as a result.[61] And don't forget, remember to look straight at the camera.

## Are oysters really the food of love?

That is the popular myth, but is it really true? And, if it is, why should that be the case? Is it to do with the sensory properties of oysters, or the symbolism associated with something that resembles the female sex? M. F. K. Fisher, the famous food writer, was certainly convinced that this bivalve was a popular aphrodisiac because of its 'odour, its consistency, and probably its strangeness'.[62] Or is this recommendation perhaps nothing more than an amalgam of tradition, folklore and old wives' tale?

Could it be something to do with their expense? Men, after all, have been shown to opt for more expensive foods when in the presence of an attractive female.[63] Well, this

*Lunch with Oysters and Wine* by Frans van Mieris.

---

* In fact, this seemed like such a good idea that, together with a couple of colleagues, we even created a program that enabled anyone to upload a few pictures of themselves and pay people online just a few cents apiece to assess which picture shows you at your best.

would seem unlikely, at least in a historical context, given that oysters were once one of the cheapest foods you could buy. The suggested link between oysters and passion dates back at least as far as the ancient Greeks, and one finds it in later paintings too, such as *Lunch with Oysters and Wine* by Frans van Mieris (1635–81). No doubts about what this lascivious-looking chap has in mind. My guess is that oysters may have been the best bet before Lynx came along!

Casanova also trusted in the seductive power of oysters. Taking no chances, he would regularly down fifty for breakfast before a hard day's lovemaking. Unfortunately, as far as I am aware, no one has yet performed the relevant 'randomized controlled trial' experiment in order to determine whether regular oyster eating really does enhance one's seductive success. Anecdotally, the connection was the zinc that is found in both oysters and semen. However, the closest to rigorous, or should that be vigorous, research so far on this point comes from a group of Italian scientists showing that oysters are high in a couple of unusual amino acids. Crucially, cooking reduces the acids' concentration, hence the recommendation to eat them raw. Injecting these compounds into male rats results in an increased production of testosterone. Another paper, meanwhile, documented an increased sperm count in, wait for it, rabbit bucks. That said, it is a rather big leap to jump from frisky rats and randy rabbits to romance in humans. And, anyway, even if the general claim were true, then there is always the risk of food poisoning to contend with. According to one survey, 70 per cent of all British oysters are potentially contaminated with norovirus. I would imagine that would put the dampers on any mollusc-induced fit of passion, no matter how many you had eaten.

It is often said that men find women more attractive if they (the men, that is) drink beer, and lots of it. Certainly they tend to find women more attractive the closer it gets to closing time.[64]

This is the so-called 'beer goggles' effect, and a number of psychologists with nothing better to do have assessed its veracity.[65] The effect proves to be real enough; women really do look more attractive to men who are under the influence, especially those women at the plainer end of the spectrum. Interestingly, the uplift is not specific to women; beer goggles have been shown to enhance the appearance of modestly attractive landscape paintings too.[66] But before you order that next 'beauty-enhancing' drink, let me just remind you of Porter's comment to Macduff in *Macbeth* that drink 'provokes the desire, but it takes away the performance'.[67]

## *The voice of desire: is beauty really in the ear of the beholder?*

Have you ever spoken to someone over the phone who had such a lovely voice that you instantly found yourself wanting to meet them in person, imagining just how attractive they must be? Does that even make sense? Can we meaningfully infer anything about what someone looks like merely from the sound of their voice? Or, to put it another way, are judgements of physical attraction really correlated across the senses?

For women, it makes sense, given that testosterone levels affect the male voice as well as changing the morphology of the human face. Think of how a boy's voice drops at puberty, and analysis of hunter-gatherer groups has shown that those with a lower voice tend to have more reproductive success.[68] Women's judgements of masculinity based on the sound of a man's voice are correlated with ratings of their face when made independently (though contraceptive use can interfere with such judgements).[69] The sound of naturally cycling women's voices also changes subtly as a function of where they happen to be in the menstrual cycle.[70]

Intriguingly, the males of many species actually lower the

pitch of their call or growl in order to make themselves appear larger when in a confrontational or competitive mating situation.[71] Small dogs also pee higher when they cock their legs to leave the same impression.[72] These are both examples of what is known as 'dishonest signalling'. Perhaps it is the fact that any single sensory cue can be so easily modified to change how it is perceived that helps explain why we are better off relying on multisensory judgements. They are simply more reliable (that is, it is simply much harder to 'fake it' in several senses simultaneously).

## Welcome to the smell dating agency

Some years ago, a Glasgow-based artist by the name of Clara Ursitti spent six months embedded at the Crossmodal Research Laboratory. Before coming to Oxford, Clara had created various scent-based art installations, often involving the synthesized smell of semen in all manner of unusual places. This perhaps explains one of Clara's favourite party tricks when she was with us. While the lab was out having dinner at a restaurant Clara would whip out a bottle of the stuff and waft it around surreptitiously. There were some very funny looks exchanged between the diners at the other tables, I can tell you. Fortunately, people are typically very poor when it comes to localizing the source of ambient odours, hence no one ever managed to figure out where that strange smell was coming from.

In one of her other projects, Clara had a group of people wash with unscented products for a few days while wearing the same T-shirt. The shirts were then bagged and the owners invited to come along and sniff out a few potential hot dates/mates. Just think of it as speed dating for your nose. People took it in turns to smell the shirts and pick out three whose owners they would most like to hook up with. The event went by the name of Pheromone Café. There were certainly some surprises when the owners of those oh-so-desirable body odours were identified, especially for those heterosexual males who had picked out T-shirts belonging to other hairy blokes, and not the delectable lady that they had perhaps been imagining.

Even if someone looks irresistible, their attraction can be undermined by olfactory evidence. A friend of mine used to go out with a stunningly beautiful woman whose body odour just wasn't quite right. What to do? Ultimately, I think that you have to go with the nose as the final arbiter, but visual dominance can be hard to override, especially for men.

## *Multisensory magic: I love you with all my senses*

Putting what we have seen in the last few sections together, it should by now be clear how each of the senses, be it sight, smell or sound,* provides useful information about the mating potential of a significant other.[73] Key cues to attractiveness come from facial and bodily appearance, voice and smell. That said, men and women differ somewhat in terms of the sensory cues that they prioritize, perhaps because of their differing goals as far as mate selection is concerned.

What is currently a little less clear is which evolutionary account does the best job of explaining the adaptive benefits of integrating the various cues to attraction. According to one version, each of our senses provides a relatively independent cue, or source of information about the 'fitness' in an evolutionary sense, of a potential mate – think here of the earlier-mentioned major

---

* Touch plays a very important role in affiliative bonding too, though there isn't space to get into that here.

histocompatibility complex, which can be smelled but not heard or seen. This is what is known as the multiple messages hypothesis. Alternatively, however, the senses are thought to convey information that is at least partly redundant – think about the correlated auditory and visual cues to masculinity that we came across earlier. This is known, unsurprisingly, as the redundant signals hypothesis. Meanwhile, according to a third account, a person's genetic quality is expressed by combining various phenotypic* traits that individually indicate health and fertility.[74] Complicating matters somewhat, these theories aren't necessarily mutually exclusive, and the degree to which any of them applies may depend on the particular trait or attribute being assessed. Nevertheless, by combining the cues provided by both proximal senses (smell) that tell us what is happening close by with the information from the distal senses (e.g. vision and hearing), it becomes possible to detect positive traits both from a distance (voice and visual appearance) as well as up close (body odour).

While it may be relatively easy for us, or other creatures, to hack any one of the senses – think make-up for vision, fragranced products to mask body odour, or men lowering the pitch of their voices to sound a little more macho – it is going to be much harder to modify all (or several) of these cues simultaneously.

Ultimately, though, any one of us can hack the senses in order to optimize our multisensory appeal to others. While it is partly a matter of natural endowment, partly self-belief and partly a healthy dose of genetics, the evidence shows that we can all increase our appeal to others by making the most of what our senses have to offer.[75] The benefits of working with all your senses are potentially better than Botox . . . and without the

---

* An individual's phenotype refers to their observable characteristics, such as their morphology and behaviour, that are thought to result from the interaction of his or her genotype with the environment.

need for fillers either! So, next time you upload some pictures onto that dating site, make sure to smile while looking directly into the camera. Holding a guitar probably won't harm your chances either. I'd also recommend getting the pictures you upload peer reviewed, while thinking carefully about how much red you show. And as soon as you get the chance, make sure to spray some pleasant fragrance into the air.

So, having taken a trip around the sense hacking of everyday life, all that remains for me to do is to summarize the key themes that have cropped up repeatedly in the text so far, and discuss what the future may hold so far as sensehacking is concerned.

# 11. Coming to Our Senses

What does the future hold for sensehacking? In the preceding chapters we have learned how our growing understanding of the senses, and the manifold connections that exist between them, is already enabling some of the most innovative individuals and organizations to hack both their senses and ours. Sensehacking is now a fact of life, no matter whether we happen to be in the car, at work or out shopping. In the years to come, I believe that it will be increasingly common in medical settings, and even in the privacy of our own homes.

Many of the latest sensehacking insights are already being used to help us to sleep better without the need for pills, to eat less without feeling hungry, to look our best without plastic surgery, to drive more safely and to get the most out of our exercise regimes without it always feeling like such a chore. All this, and much, much, more can be achieved simply by paying attention to the often silent influences that our senses have over us. The more we train, or educate, our senses, the more we stand to gain from life.[1] Who amongst us would not benefit from taking better care of their senses? So, why not 'Indulge your senses', as the seductive sensory strapline from a 2001 marketing campaign for Fairy washing-up liquid exhorted.

Sensing and sensation are fundamentally pleasurable.[2] If, for whatever reason, you have any doubts about this, then all you need to do is watch as a congenitally deaf child has their new cochlear implant switched on to be immediately convinced.* Watch as they burst into tears of unmediated joy on hearing for the very first time. There are many heart-warming examples of the spontaneous eruption of emotion online, that almost primal rush of pleasure that is so often associated with experiencing a new channel of sensation, be it the sensehacking of sight or sound.

## Sensory deprivation

What do you think is one of the CIA's favourite legal forms of torture? You got it – sensory deprivation. While raw sensation can undoubtedly be intensely pleasurable, its enforced absence, sensory deprivation, can be absolute torture. Think of it as sensehacking, but not in a good way. Eyes blindfolded, ears blocked, smell, taste and tactile sensation all minimized. While this denial of external stimulation leaves no physical scars, the psychological

Many 'high value' detainees were subjected to sensory deprivation at Guantánamo Bay. Here convicted terror-plotter José Padilla is wearing blacked-out goggles and earmuffs – basic deprivation tools intended to soften prisoners up mentally by plunging them into a sensory void.

The idea of punishing criminals by restricting their access to sensation actually dates back much further, first being suggested in an 1846 treatise by German doctor Ludwig Froriep.†

---

\* E.g. www.youtube.com/watch?v=ZLRhGUhxKrQ or www.youtube.com/watch?v=yZ6vSn7PaPI.
† Jütte (2005); Salon, 7 June 2007, www.salon.com/2007/06/07/sensory_deprivation/.

damage may be permanent. After no more than a couple of days, people start experiencing rich and vivid visual and auditory hallucinations, presumably to make up for the lack of external stimulation. This often leads to psychosis, followed closely thereafter by complete mental breakdown. The consequences can be so severe that there are those who question whether this practice contravenes the Geneva Conventions.

On the positive side, plenty of alternative treatment centres are offering those suffering from the sensory overload of modern life the opportunity to undergo a little voluntary sensory deprivation. The idea is that we can recharge our senses simply by floating for a while in absolute silence and total darkness in a tank of water heated to body temperature. Don't stay too long in there, though; the brain abhors a vacuum and soon enough the sensory hallucinations will start. We prefer sensation to be externally generated, but if, for whatever reason, that is not possible then our brains will readily step up to the plate and provide it for us, sometimes within hours.[3]

## Are you suffering from sensory overload?

While sensory deprivation can undoubtedly be used to torture, its opposite, sensory overload, can be an equal torment. Indeed, many of us spend far too much of our time distracted by technology, suffering from the symptoms of sensory overload. You know that things have reached a tipping point when even the tech titans of Silicon Valley say they have finally had enough. Apparently, there is now an increasingly popular trend of trying to avoid 'sensation overload' by means of what is known as 'dopamine fasting'.*[4]

---

* Though, as one commentator notes, the term 'dopamine fasting' is actually a bit of a misnomer. It's more of a stimulation fast, really.

This involves a short-term withdrawal from all forms of social contact. Direct eye contact, for instance, is to be avoided at all costs as it is simply far too arousing. At the same time, all other forms of stimulation, such as those provided by eating and drinking, are restricted too. The hope is that the eventual return of sensation, for those who have chosen to take this most abstemious of paths, will then be all the more pleasurable, as well as easier to manage.[5] Well, they do say that absence makes the heart grow fonder, don't they? And if the tech entrepreneurs are struggling to cope, you can imagine what 'a visual and auditory assault on the senses' daily life can be for those with special needs, or for the growing number of people with some recognized form of sensory-processing disorder.[6]

Our addiction to handheld technologies undoubtedly plays its part in terms of maintaining sensory overload too. After all, many of us complain about there being too much information, as we try to multitask using all that technology has to offer us.[7] At the same time, however, the all-too-rapid rise of megacities (those whose populations exceed 10 million) across the globe can only be making matters worse. Stanley Milgram, the Yale psychologist famous for conducting experiments on obedience in the 1960s,* certainly thought so, writing this in 1970: 'City life, as we experience it, constitutes a continuous set of encounters with overload.'[8] Never mind *Sex and the City*, it is more a case of Sensory Overload and the City, if you ask me. What do they promise other than more noise, more pollution, the restorative sights, sounds and smells of nature nothing but a distant memory, perhaps fondly recalled while watching the latest

---

* Milgram conducted research showing that people could all too easily be convinced to apply increasingly severe electric shocks to another person when instructed to do so by an authority figure. In fact, no shocks were ever administered. It was just an actor convincingly simulating the pain. However, the participants who took part in the study were not to know that.

doctored nature show on TV? The contemporary shift in the 'sensorium' that we have created for ourselves is, make no mistake about it, causing problems for our health, for our well-being and for our cognitive function too.[9]

Optimizing the multisensory environments that surround us has never been more important, especially given that those of us living an urban existence spend something like 95 per cent of our lives indoors. As of 2010, more people around the globe lived in cities than in rural areas.[10] And, according to the latest projections from the United Nations, 68 per cent of the world's population is expected to live in urban areas by the year 2050, many in megacities.[11] No matter whether we are talking about work, home, health, exercise or sleep, the fact of the matter is that the negative consequences of the unnatural indoor environment that we have fashioned for ourselves, removed as it is from most sources of natural light, not to mention continually exposed to a host of airborne pollutants and noise, are there for all to see (and smell, and hear).[12] The incidence of seasonal affective disorder and sick building syndrome continues to be higher than any of us would like.[13] At the same time, the chronic sleep problems that currently blight so many of our lives have also been attributed, at least in part, to the poor diet of sensory stimulation that we nourish ourselves with.[14] All that before we get to the obesogenic environments in which we find ourselves, which, when combined with our increasingly sedentary lifestyles, are also partly to blame for the growing global obesity crisis.

## *Primal pleasures: the nature effect*

Seemingly every week a new study comes out showing just how beneficial contact with nature, and that includes its sights, sounds and smells, is for our health and mental well-being. The 'nature effect' is real, and we would all do well to remember it.[15] In fact, the pull of those sensory triggers that are somehow primally reminiscent of our earlier human history proves so powerful that, you will recall, we end up setting the temperature and humidity in our own homes to match the climate of the Ethiopian Highlands where humankind evolved some 5 million years ago.[16] The hope, then, is that by hacking the senses, for example by recognizing that our thermal comfort is partly determined by what we see, and not just by the ambient temperature indoors, we may all be able to design multisensory environments and interventions that, for example, help us to tackle global issues like climate change by using warm colours to reduce our energy use.[17]

## *Welcome to the sensory marketing explosion*

As we have seen several times throughout this book, sensory marketers are partly to blame for the sensory overload that affects so many of us.[18] For years they have been vying for our attention in store, using all of the sensory touch points at their disposal. Now you know what to look for, you will see just how prevalent this is: the plethora of adverts these days for products and experiences promising to do everything from seducing your senses to refreshing them, not to mention stimulating, exciting and intoxicating them to boot. In their drive to engage all of your sensory faculties, the marketers have left no verb unturned, or so it would seem.

That said, all too often they have failed to deliver on their promise. In my opinion, while they often talk the talk about engaging our senses, too often they fail to deliver the right balance of sensory stimulation. I speak from experience here, having been the Global Head of Sensory Marketing at the JWT Advertising Agency for a number of years.[19] Nevertheless, the marketers' intuitive drive to want to tap *all* our senses is understandable, given the emerging neuroscientific evidence from the rapidly growing field of multisensory perception research.[20] After all, what are life's most pleasurable experiences if not intensely multisensory.*

Who can blame all those companies who are already using the power of the senses to lure us into their stores, before nudging us in this direction or that? I am the first to admit that such multisensory manipulation in the marketplace raises ethical questions, especially given the fact that we are so rarely aware of these sensory triggers, not to mention their impact on pretty much every aspect of our daily lives. The growing realization of just how much power the senses have over us, both individually and, more importantly, when combined (which, come to think of it, they nearly always are), can be scary. However, once you recognize their unquestionable influence over us, there really is no going back. That is, there is no good reason, at least as far as I can see, why we should not all use the power of the senses to our own advantage, sensehacking our own experience as well as that of those around us whom we love. You can think of it as 'sensory nudging', at least when the goal of sensehacking is societal good.[21] And talking of using the senses for well-being . . .

## Sensism: a mindful approach to the senses

We all need to pay more attention to our senses than we do at present. You can call it a mindful approach to the senses, if you

---

* Just think about food and sex, should you be struggling for inspiration.

want, though my preferred terminology, coined in an industry report published almost twenty years ago, is 'Sensism'.[22] Sensism is fundamentally about providing a key to improved well-being by considering the senses holistically, understanding how they interact and incorporating that understanding into our everyday lives. While the sensory marketers may well have been in the vanguard as far as curating our sensory experience is concerned, that certainly does not mean that they have all the answers, far from it. Just take the following quote from Philip Kotler, one of the leading marketers of the last half century. As he put it in a review article on marketing in the third millennium, 'The senses are the vehicles by which we experience the world, but the question remains, what does it mean to say that "we experience" something, and that the experience was a satisfying one or not?'[23]

One of the fundamental problems faced by many marketers is that, while they have unquestionably recognized the power of

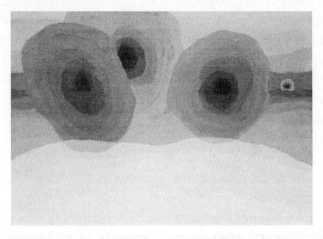

Arthur Dove's *Fog Horns* (1929). The loud, low-pitched warning sounds are conveyed almost synaesthetically by the size and darkness of the circles in the painting.

the senses when targeted individually, they have pretty much all failed to grasp just how much interaction there is between them. And, as we have seen throughout this book, perception is nothing if not multisensory. So often, what we see is affected by what we hear, what we smell by what we feel, what we feel by what we see, and so it goes on. Sensehacking is built on our growing understanding of such sensory interconnections. In recent years, advertisers have also started to communicate with us by sensehacking these sometimes surprising, but nevertheless widely shared, connections, or correspondences, between the senses.[24] This is something that composers, artists and designers have been doing intuitively for centuries; just think of Kandinsky, Scriabin or the lesser-known Arthur Dove.[25]

## Do you have the sensory balance right?

Ultimately, sensism is all about finding the right balance of sensory stimulation in our lives. We need to recognize that the sensory overload so many of us complain about is really only affecting our higher rational senses, namely our eyes and ears. As we have seen, too many of us are suffering from a neglect of our more emotional senses – namely touch, taste and smell.[26] There are those like Dr Tiffany Field in Florida who have been arguing for decades that we, as a society, especially in the West, are suffering from what she calls 'touch hunger', be that in the home or in healthcare.

We have ignored the skin for too long. There has been an almost wilful failure to recognize just how important it is to stimulate our largest sense organ.[27] The suggestion that we need to stimulate the skin, be it through aromatherapy massage or a lover's caress, was for too long seen as being somehow unscientific. Nevertheless, the latest developments in social, cognitive and affective neuroscience are increasingly highlighting the profoundly beneficial effects of stroking the skin. Benefits, moreover, that extend all the way from a baby's earliest encounter with the world through to helping make up for the sensory underload experienced by many old people today, whose wrinkly skin people are less willing to touch.[28] So, as scientists uncover the optimal stimulation parameters, isn't it time that we all started to sensehack our skin, or rather our sense of touch, a little more intelligently?

We all like different amounts of sensory stimulation, with some – the sensation seekers, or 'sensory junkies' – craving it while others prefer much less.[29] We all live in our own worlds of sense. Ultimately, there is no right or wrong in terms of the optimal amount of stimulation we receive, just a rich world of variety. In this regard, it is heartening to see how the field of sensory design is coming to recognize the gradations of sensation we like and/or can deal with. The good news is that across a range of situations, this is leading to a more accessible and inclusive approach to the sensehacking of design, one that recognizes the different sensory worlds in which we all live.[30] Just remember that your experience of the world is going to be subtly (or not so subtly) different from everyone else's.

We should, I think, be questioning the dominance, or hegemony of the visual in contemporary society. We should all be asking whether this particular hierarchy of the senses that we find ourselves with currently is the right one, be it for ourselves or for the societies in which we live. Looking across cultures and history, one finds that there are and, more importantly, always have been many different possible sensory hierarchies.[31] Just think about it for a moment: it really is not that long since we would tell the time not by looking at our watch but by listening to the chimes of the church clock, and when flowers and fruits were grown for their fragrance and flavour rather than their size and uniformity.[32] I know which I prefer, and I think you do too (this just leaving the question of why the supermarkets do not give us what so many of us say that we want). So, which balance of sensory stimulation do we want, as individuals, and for the societies in which we live? These are precisely the kinds of questions that we should all be asking ourselves.

## Social isolation in the pandemic era

This sensory imbalance, I fear, is only going to get worse as social distancing and long-term isolation as a result of the coronavirus pandemic, and indeed any others that might follow it, take their inevitable toll on our emotional well-being.[33] It is in this context that I am particularly interested by ideas around digital tactile stimulation as a means of making up for 'touch hunger'. If it were possible to send a caress, or hug, to a loved one at a distance, over the internet, say, could some of the social isolation that so many of us are suffering from be ameliorated? There have long been items of clothing that can transmit tactile stimulation from one person to another, and the so-called Hug Shirt was listed by *Time* magazine as one of the best inventions of 2006. Or perhaps what you are looking for in these most taxing of times is the rather more reassuringly firm grip of something like the HuggieBot, a modified 450lb (204kg) research robot?[34]

In the 1950s Harry Harlow conducted a series of shocking experiments showing that socially isolated baby monkeys preferred to hold on to the furry-textured 'mother' rather than the wire-monkey that provided food. In other words, they preferred sensation to sustenance.

The problem, though, as we saw in the Healthcare chapter, is that this kind of digital, mechanical, mediated touch simply does not convey the same social, emotional or cognitive benefits as actual interpersonal contact. Perhaps it is nothing more than a matter of getting the temperature right. For while human touch is typically warm, mechanical touch tends to be thermally neutral or cold. Or it may be that we need the right smell, possibly involving some of those chemosensory pheromonal signals that we learned about in the Dating chapter. Or, then again, maybe there are some things that we simply can't simulate, like the emotional connection that lies behind a caring interpersonal caress. Just 'going through the motions' may not be good enough, at least not in this case. Ultimately, we need to recognize that touch is not merely a matter of what happens on the skin surface. Our experience of touch, and of being touched, is influenced by what we see, hear and smell. It is, then, like pretty much everything else in life, a multisensory phenomenon.[35] It is only by acknowledging the multisensory nature of perception that we can hope to hack our senses successfully.

You may remember that we came across much the same issue concerning the mediated nature of sensation earlier in the book when looking at whether plastic plants were as good as the real

The Hug Shirt from Cute Circuit contains ten actuators and uses Bluetooth technology to deliver 'personal' touch over a distance.

thing, and whether a video-feed of nature was as good as a room with a view.³⁶ In the pandemic era, I suspect that many of the same issues remain relevant for anyone trying to maintain social relations in isolation by means of internet-based digital commensality.³⁷

## Sensehacking new sensations

While the majority of the sensehacking examples in this book have involved the five main senses, there are a small, but growing number of biohackers who would like to go beyond what nature has provided them with. For example, Catalan avant-garde artist and cyborg activist (so not the kind of person you meet every day) Moon Ribas has a seismic sense. An implant in her arm

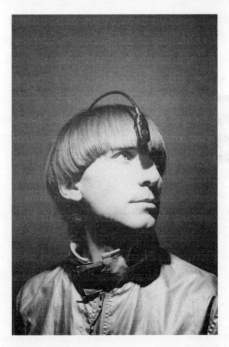

Human cyborg, or should that be 'eyeborg', Neil Harbisson.

allows her to feel the earth's seismic activity, with a sensor in her elbow vibrating every time an earthquake occurs anywhere on the planet.[38] And there is also Neil Harbisson, a British artist who grew up in Catalonia and was born with achromatopsia, a severe form of colour blindness. He calls himself a human cyborg, or 'eyeborg', and may well be the only person whose body-augmentation device is actually visible in his passport photo. He has a chip that translates the colour picked up by a camera surgically attached to his scalp into vibration and sound.* Note that this counts as an example of sensory substitution rather than sensehacking, in that it allows the artist to 'hear' colours, including those light waves at the infrared and ultraviolet ends of the spectrum, which the rest of us simply cannot sense. According to Harbisson, his antenna should not be thought of as an add-on technology, but rather a sense organ in its own right.[39]

If you like the idea of experiencing a new sense for yourself, then London-based company Cyborg Nest may be just the one for you. They are currently at the forefront of what sensehacking has to offer, and have been selling their North Sense device commercially for several years. The basic idea is that a 1-inch/2.5mm square body-compatible silicon-encased magnet attached to your chest vibrates briefly whenever you are facing geomagnetic north. The device, which sits outside of the body, is attached by means of a pair of titanium barbell piercings inserted under the skin. Now, if that sounds invasive, well, that's kind of the idea. So, will the widespread availability of devices such as the North Sense one day prompt a 'cyborg' evolution in human sensory abilities?

According to Cyborg Nest's founders, Scott Cohen and Liviu Babitz, they have already shipped several hundred units to a

---

* You can imagine the scene, though, when by chance I happened to find myself behind him at airport security once . . . All hell broke loose!

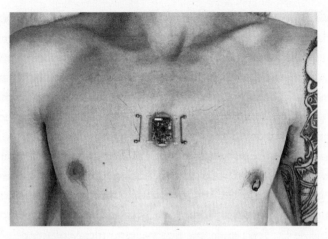

Cyborg Nest's North Sense. (Nipple ring optional.)

wide variety of open-minded individuals from around the world since their device launched in 2017. Babitz, the company's CEO, certainly has some grand ideas about what might follow should it actually prove possible to give people an extra sense. In one interview, he even went so far as to say, 'Everything we ever created, we created because we have senses. If we have more senses, we are lifting the creation glass ceiling of humanity exponentially higher.'[40]

Now, quite what the experience of a north sense is like, and how it changes over time wearing such a device is, at present, unknown. That said, the hope expressed by one IT geek from Novosibirsk in Russia who had one of these devices fitted was absolutely clear: 'I think at some point it will grow into a feeling, I will stop feeling vibration and will feel direction instead.' He goes on to say that one day he believes that it really would become his sixth sense. In fact, so enthusiastic was this biohacker that he later went on to say that he can't wait to implant one in his wife! Let's just hope he asks for her permission first. Every bionic man needs his bionic woman, it would seem.[41] Ultimately, though, it is still unclear whether any of the implants and/or attachments favoured by the biohackers really do give their users a 'new sense', as the more ardent grinders and transhumanists* would have us believe.[42] Nevertheless, their attempt to challenge the limits of what sensation might be (or one day become) shines a light on what the future of sensehacking might one day become.

And finally here, looking even further into the future, there are those who would rather bypass the senses and body entirely and go straight to the control centre – the brain. As a sign of things to come, in 2017 Elon Musk, the billionaire entrepreneur behind both Tesla and SpaceX, started a new company called Neuralink. This California-registered corporation is working on implanting a tiny chip directly into the brain.[43] Called 'neural lace', the idea is one day to enable people to upload or download their thoughts to, or from, a computer. The hope is that such products, should they ever become commercially viable, could allow humans to achieve higher levels of cognitive function.† And, who knows, one day such brain implants may even be used to sensehack our perception of reality, as was captured in the hit Hollywood film *The Matrix* more than twenty years ago:

> If the virtual reality apparatus, as you called it, was wired to all of your senses and controlled them completely, would you be able to tell the difference between the virtual world and the real world?

---

* Transhumanism is the name of a movement based on the belief that the human body can be improved by scientific advancement.
† Facebook is apparently also exploring similar technology at Building 8, its secretive hardware division. The group has reportedly been developing non-invasive brain–computer interface technology that would allow people to communicate with external hardware devices.

> What is real? How do you define real? If you're talking about your senses, what you feel, taste, smell, or see, then all you're talking about are electrical signals interpreted by your brain.
>
> From *The Matrix* (1999)

## Coming to our senses

Looking to the future, I am especially excited to see whether sensehacking can one day be harnessed not just to modify the 'mundane' experiences of everyday life, but rather to deliver the extraordinary ones, those that have the potential to be truly transformative. I am also eager to know whether the biohackers and grinders are able to enable new senses, such as a magnetic sense, a seismic sense, or something even more bizarre, by sensehacking from within. Most likely we will only come to a transformative relationship with our senses by further growing our understanding of them, paired with a recognition of the unique worlds of sense that each of us inhabits. Achieving that goal will no doubt be helped by the emerging science of the senses, and the multitude of interactions that exist between them.

Ultimately, the future of sensehacking will also be facilitated by the artists, architects and designers who will ideally transform the science of the senses into intriguing and engaging multisensory experiences, and question the status quo as far as the hierarchy of the senses is concerned. After all, given how much time we urbanites spend indoors, even when not in lockdown, architects and urban planners clearly have a crucial role to play in helping to deliver the kinds of multisensory environments that can help to promote our social, cognitive and emotional well-being.[44] At the same time, however, sensehacking will undoubtedly also benefit from the contextualization of the senses, and the mediated sensorium, that is offered by the growing body of historians, anthropologists and sociologists of the senses who, along with many others, have so recently taken the 'sensory turn'.[45]

Now that I have shared some of the secrets of sensehacking with you, I hope that you will be inspired to make the most of what your senses have to offer, be it in terms of enhancing your social, emotional or cognitive well-being. At the same time, knowing more about how your senses affect you should make it that bit harder for the wily marketers out there to hack them.

# Appendix: Simple Sensehacks

Towels that smell nice *feel* softer (page 26).

Food tastes 10 per cent better if you use a tablecloth – you'll also eat 50 per cent more of it (pages 34–5).

If you like showers, cold ones reduce sick leave by 29 per cent (to put this figure into perspective, regular exercise reduces it by 35 per cent) (page 40).

It's the relaxing scent of face cream that (temporarily) removes wrinkles (page 41).

The sound of nature is relaxing (no surprise there), and the more bird species you can hear, the more relaxing it is (pages 51–2).

Noisy neighbours? Listen the same thing as them and you'll sleep better (page 73).

Struggling to sleep and only have one earplug? You should put it into your right ear (pages 79–80).

If you like baths, the ideal water temperature to help you sleep afterwards is 40–42.5°C (104–108.5°F) (page 75).

Family cars peacock in 'sport' modes by showing red lights and increasing engine noise – the performance of the car is very often unaffected (page 96).

Indoor plants can reduce office air pollution by 25 per cent and cleaner air increases productivity by 8–11 per cent (page 134).

Women are often cold in offices because their metabolism is lower. Increasing the ambient temperature results in a decrease in men's performance of 0.6 per cent per degree, while increasing female performance by 1–2 per cent, so turn the temperature up (page 122).

Had a stressful meeting at work? Smell a different scent to mentally reset (page 124).

On average, people lose eighty-six minutes a day due to disturbance when working in an open-plan office. If you can't work from home, background music increases productivity by 10–20 per cent (pages 124, 129).

When you smell baked bread in a shop or fast food outlet, despite common opinion, it is likely the real deal (page 145).

Shoppers spend 38–50 per cent more when slow as opposed to fast music is played (pages 150, 165).

Want to exercise harder? Increase the musical tempo by 10 per cent – and you'll also enjoy it more (page 207).

Want a competitive edge in tennis? Grunting really does help (page 210).

The noise of the crowd affects how likely a referee is to hand out a yellow card – so shout louder (page 211).

If you smile when exercising, you can improve the economy of your running by more than 2 per cent (page 213).

If you simply *taste* carbohydrate for a few seconds once every 7–8 minutes or so when exercising (e.g. by putting a sports drink in your mouth and then spitting it out) sporting performance can increase by 2–3 per cent (page 213).

Choosing a kit colour for your sports team? Black will make you more successful (pages 216–17).

Taking a date to the movies? Increase your chances of it ending well by watching a thriller (page 224).

You can smell a person's age (page 231) but not their gender (pages 243–4).

# Acknowledgements

As always, my thanks go to Professor Francis McGlone, formerly of Unilever Research, now of the University of Liverpool, for funding much of the basic research on which many of the ideas in this book were built. He has been there since the beginning with his unwavering belief in the importance of touch and the other senses. Support from ICI funded the Sensism report, published in 2002, which again proved foundational to so much of my subsequent sensehacking research over the years. Since then, my thanks go to Christophe Cauvy for bringing me into the sensory marketing fold at JWT, and to Rupert Ponsonby (of R&R) for making sure that my throat was never parched by providing a range of most intriguing multisensory libations. Many of the ideas around sensehacking healthcare came directly out of discussion and ongoing collaboration with Steve Keller, now working at Pandora. Thanks also to chef Jozef Youssef of Kitchen Theory for demonstrating the power of sensehacking in so many of the delicious dishes served at his chef's table in High Barnet. The late Jon Driver also deserves a heartfelt acknowledgement for shepherding me into a life of research in the first place (and saving me from the City). If it hadn't been for his broken television, not to mention his academic acumen and rigour, my life would likely have turned out very differently. I have also benefited immensely from the support of many gifted students and academic collaborators over the years.

Daniel Crewe and Connor Brown at Penguin have also patiently supported the development of this work since its inception some many years ago now! And finally, to my wife Barbara (Babis) for her critical yet constructive support every step of the way.

# Notes

## 1. Introduction

1 Galton (1883), p. 27.
2 Bellak (1975); Malhotra (1984).
3 www.accenture.com/_acnmedia/accenture/conversionassets/microsites/documents17/accenture-digital-video-connected-consumer.pdf.
4 Colvile (2017).
5 Spence (2002).
6 Montagu (1971).
7 Classen (2012); Denworth (2015); Field (2001); Gallace and Spence (2014).
8 Cohen et al. (2015); Goldstein et al. (2017).
9 Sekuler and Blake (1987); US Senate Special Committee on Aging (1985–6), pp. 8–28.
10 Classen et al. (1994); Herz (2007); *Touching the rock: An experience of blindness.* London: Society for Promoting Christian Knowledge, www.brighamsuicideprevention.org/single-post/2016/05/08/Paving-the-path-to-a-brighter-future.
11 *Financial Times*, 4 June 2013, 1; *New Yorker*, 26 October 2012, www.newyorker.com/magazine/2015/11/02/accounting-for-taste.
12 See www.johnsonsbaby.co.uk/healthcare-professionals/science-senses.
13 Ho and Spence (2008); Spence (2012a).
14 *Businesswire*, 27 July 2015; www.businesswire.com/news/home/20150727005524/en/Research-Markets-Global-Cosmetics-Market-2015-2020-Market.

15 *Guardian*, 30 October 2017, www.theguardian.com/lifeandstyle/2017/oct/30/sad-winter-depression-seasonal-affective-disorder; Ott and Roberts (1998). For my early work with Dulux paints and Quest fragrances on the design of more productive and healthier indoor environments see Spence (2002) and, for the latest review, Spence (2020f).
16 Adam (2018); Huxley (1954); Walker (2018).
17 Cutting (2006); Monahan et al. (2000); Kunst-Wilson and Zajonc (1980).
18 Hepper (1988); Schaal and Durand (2012); Schaal et al. (2000).
19 Hoehl et al. (2017); LoBue (2014).
20 Dobzhansky (1973).
21 Batra et al. (2016); *New York Times*, 16 May 2014, www.nytimes.com/2014/05/17/sunday-review/the-eyes-have-it.html.
22 Karim et al. (2017); *New York Times*, 27 November 2008, B3, www.nytimes.com/2008/11/28/business/media/28adco.html.
23 Salgado-Montejo et al. (2015); Wallace (2015); Windhager et al. (2008).
24 Spence (2020c).
25 Sheldon and Arens (1932).
26 Croy et al. (2015); Field et al. (2008).
27 Cheskin and Ward (1948); Martin (2013); Packard (1957); Samuel (2010).
28 Fisk (2000); Spence (2002).
29 Gori et al. (2008); Raymond (2000).
30 Hutmacher (2019); Meijer et al. (2019).
31 Howes (2014); Howes and Classen (2014); Hutmacher (2019); Schwartzman (2011).
32 McGurk and MacDonald (1976).
33 Wang and Spence (2019).

## 2. Home

1. Dalton and Wysocki (1996); *Financial Times*, 3 February 2008 (House and Home), 1.
2. Glass et al. (2014); Spence (2003); Weber and Heuberger (2008).
3. *Independent*, 14 May 2018, www.independent.co.uk/news/long_reads/sick-building-syndrome-treatment-finland-health-mould-nocebo-a8323736.html.
4. Quoted in Corbin (1986), p. 169.
5. *Crafts Report*, April 1997, https://web.archive.org/web/20061020170908/www.craftsreport.com/april97/aroma.html; *Ideal Home*, 15 March 2018, www.idealhome.co.uk/news/smells-sell-your-home-scents-197937; McCooey (2008); *The Times*, 19 March 2014, 5.
6. Haviland-Jones et al. (2005); Huss et al. (2018).
7. Baron (1997); Holland et al. (2005).
8. Herz (2009).
9. Haehner et al. (2017); Spence (2002).
10. Le Corbusier (1948).
11. Spence (2020e).
12. Fich et al. (2014).
13. Clifford (1985); McCooey (2008).
14. Appleton (1975), p. 66; Manaker (1996).
15. Dazkir and Read (2012); Thömmes and Hübner (2018); Vartanian et al. (2013).
16. Lee (2018), p. 142; McCandless (2011).
17. Zhu and Argo (2013).
18. 'Music makes it home', http://musicmakesithome.com, in Lee (2018), p. 253.
19. Spence et al. (2019b).
20. Baird et al. (1978); Meyers-Levy and Zhu (2007); Vartanian et al. (2015).

21 Oberfeld et al. (2010).
22 Bailly Dunne and Sears (1998), p. 3; Crawford (1997); *New York Times International Edition*, 31 August – 1 September 2019, 13; http://antaresbarcelona.com.
23 Pallasmaa (1996).
24 Etzi et al. (2014); Demattè et al. (2006).
25 Imschloss and Kuehnl (2019).
26 Itten and Birren (1970); Le Corbusier (1972), p. 115; Le Corbusier (1987), p. 188; Wigley (1995), pp. 3–8.
27 Küller et al. (2006); Kwallek et al. (1996).
28 Costa et al. (2018).
29 Evans (2002), p. 87; Jacobs and Hustmyer (1974); Valdez and Mehrabian (1994).
30 Quote from Oberfeld et al. (2009), p. 807; Reinoso-Carvalho et al. (2019); Spence et al. (2014a).
31 Mavrogianni et al. (2013); US Energy Information Administration (2011), www.eia.gov/consumption/residential/reports/2009/air-conditioning.php.
32 Just et al. (2019); *The Times*, 20 March 2019, 13.
33 Quoted in Steel (2008).
34 Jütte (2005), pp. 170–72.
35 Spence (2015).
36 Alter (2013); Changizi et al. (2006); *The Times*, 3 February 2017, www.thetimes.co.uk/article/think-pink-to-lose-weight-if-you-believe-hype-over-science-9rxlndnpv.
37 Genschow et al. (2015).
38 Cho et al. (2015). See also https://dishragmag.com/ (2019, Issue 2): Blue.
39 Jacquier and Giboreau (2012); Essity Hygiene and Health, 'What's your colour?' (2017), www.tork.co.uk/about/whytork/horeca/.
40 Bschaden et al. (2020); García-Segovia et al. (2015); Liu et al. (2019).

41 Watson (1971), p. 151.
42 *Smithsonian Magazine*, February 1996, 56–65; *Wall Street Journal*, 23 October 2012, https://www.wsj.com/articles/SB10001424052970203406404578074671598804116#articleTabs%3Darticle.
43 *The Times*, 26 April 2019 (Bricks and Mortar), 6.
44 Attfield (1999); Bell and Kaye (2002); Steel (2008).
45 Quote from p. 42, cited in Steel (2008), p. 197.
46 https://fermentationassociation.org/more-u-s-consumers-eating-at-home-vs-restaurant/; Spence et al. (2019-a).
47 Bailly Dunne and Sears (1998), p. 107; *Guardian*, 23 August 2017, www.theguardian.com/lifeandstyle/shortcuts/2017/aug/23/bath-or-shower-what-floats-your-boat.
48 *Daily Mail*, 19 October 2017, 19.
49 *i*, 24 March 2017, 33; Hoekstra et al. (2018); Kohara et al. (2018).
50 Buijze et al. (2016).
51 *Guardian*, 23 August 2017, www.theguardian.com/lifeandstyle/shortcuts/2017/aug/23/bath-or-shower-what-floats-your-boat; Golan and Fenko (2015).
52 Churchill et al. (2009).

## 3. Garden

1 Wilson (1984); Wilson had already won two Pulitzer prizes by the time he wrote *Biophilia*, a term he first introduced in 1979 (*New York Times Book Review*, 14 January, 43). See also Kahn (1999); Kellert and Wilson (1993); Townsend and Weerasuriya (2010); Williams (2017).
2 Treib (1995).
3 *Daily Telegraph*, 12 July 2009, www.telegraph.co.uk/news/uknews/5811433/More-than-two-million-British-homes-without-a-garden.html.

4 *Globe Newswire*, 18 April 2018, www.globenewswire.com/news-release/2018/04/18/1480986/0/en/Gardening-Reaches-an-All-Time-High.html.
5 Ambrose et al. (2020); de Bell et al. (2020).
6 Steinwald et al. (2014).
7 Glacken (1967).
8 Olmsted (1865b), available online at www.yosemite.ca.us/library/olmsted/report.html; cf. Olmsted (1865a).
9 Li (2010); Miyazaki (2018); Morimoto et al. (2006). See also Park et al. (2007).
10 E.g. Ulrich et al. (1991).
11 Louv (2005); Pretty et al. (2009).
12 Mackerron and Mourato (2013).
13 Wilson (1984); Nisbet and Zelenski (2011).
14 Kaplan (1995, 2001); Kaplan and Kaplan (1989).
15 Berman et al. (2008).
16 Knopf (1987); Ulrich et al. (1991).
17 Kühn et al. (2017).
18 Seto et al. (2012). See also Fuller and Gaston (2009).
19 Wilson and Gilbert (2005).
20 Nisbet and Zelenski (2011).
21 Ulrich (1984).
22 Moore (1981).
23 *New Yorker*, 13 May 2019, www.newyorker.com/magazine/2019/05/13/is-noise-pollution-the-next-big-public-health-crisis; Passchier-Vermeer and Passchier (2000).
24 Alvarsson et al. (2010).
25 Slabbekoorn and Ripmeester (2008).
26 Fuller et al. (2007); Ratcliffe et al. (2016).
27 Dalton (1996).
28 Hill (1915).
29 Lee and DeVore (1968), p. 3.
30 Koga and Iwasaki (2013).

31 Kaplan (1973).
32 Intriguingly, Anderson et al. (1983) found that while birdsong enhanced people's ratings of wooded, natural, and heavily vegetated urban settings, it was traffic noise (i.e. the congruent sound) that actually enhanced their ratings of the urban scenes the most.
33 Anderson et al. (1983); Benfield et al. (2010); Mace et al. (1999); Weinzimmer et al. (2014).
34 Hedblom et al. (2014).
35 Collins (1965); Romero et al. (2003).
36 Matsubayashi et al. (2014).
37 Carrus et al. (2017); Han (2007); Twedt et al. (2016).
38 Ames (1989); Frumkin (2001).
39 Seligman (1971). Wilson (1984) also has an intriguing chapter entitled 'The Serpent'; Ulrich (1993).
40 Though see Diamond (1993).
41 Hagerhall et al. (2004); Joye (2007); Redies (2007).
42 Joye and van den Berg (2011), p. 267.
43 Greene and Oliva (2009); Reber, et al. (2004); Reber, et al. (1998).
44 There are many more studies than I have space to review here. For those interested in finding out more, see Hartig et al. (2011). The results of meta-analyses stress the need for more research regarding some of the specific health claims that have been made to date: see Bowler et al. (2010).
45 Bratman et al. (2015).
46 Kabat-Zinn (2005).

## 4. Bedroom

1 Kochanek et al. (2014); *Guardian*, 24 September 2017, www.theguardian.com/lifeandstyle/2017/sep/24/why-lack-of-sleep-health-worst-enemy-matthew-walker-why-we-sleep.

2  Hafner et al. (2016); www.aviva.com/newsroom/news-releases/2017/10/Sleepless-cities-revealed-as-one-in-three-adults-suffer-from-insomnia/; www.nhs.uk/live-well/sleep-and-tiredness/why-lack-of-sleep-is-bad-for-your-health/.
3  Morin (1993); Walker (2018).
4  Hafner et al. (2016); Lamote de Grignon Pérez et al. (2019); Roenneberg (2012); Taheri et al. (2004).
5  *Guardian*, 24 September 2017, www.theguardian.com/lifeandstyle/2017/sep/24/why-lack-of-sleep-health-worst-enemy-matthew-walker-why-we-sleep; Hafner et al. (2016); Roenneberg (2013); Walker (2018).
6  Hafner et al. (2016); Understanding sleep, *Raconteur*, 4 July 2014.
7  Gibson and Shrader (2014); Sleep will never be a level playing field, *Raconteur*, 4 July 2014.
8  Harvard Medical School (2007). Twelve simple tips to improve your sleep, http://healthysleep.med.harvard.edu/healthy/getting/overcoming/tips; Wehrens et al. (2017).
9  www.nhs.uk/apps-library/sleepio/; Arbon et al. (2015); Kripke et al. (2012); Walker (2018).
10 Harvey (2003); Harvey and Payne (2002).
11 *Huffington Post*, 29 June 2015, www.huffingtonpost.co.uk/entry/smartphone-behavior-2015_n_7690448?ri18n=true.
12 Chang et al. (2015).
13 Fighting the blue light addiction, *Raconteur*, 4 July 2019.
14 Park et al. (2019).
15 *Guardian*, 21 January 2019, www.theguardian.com/lifeandstyle/2019/jan/21/social-jetlag-are-late-nights-and-chaotic-sleep-patterns-making-you-ill.
16 Chamomile tea, will you help me sleep tonight? *Office for Science and Society*, 8 March 2018, www.mcgill.ca/oss/article/health-and-nutrition/chamomile-tea-will-you-help-me-sleep-tonight.
17 Basner et al. (2014); World Health Organization (2011).
18 Arzi et al. (2012); Schreiner and Rasch (2015); Wagner et al. (2004).

19 Haghayegh et al. (2019).
20 Kräuchi et al. (1999); Maxted (2018); Muzet et al. (1984); Raymann et al. (2008); Walker (2018), pp. 278–9.
21 Chellappa et al. (2011); Czeisler et al. (1986); Lockley et al. (2006).
22 Perrault et al. (2019).
23 That, at least, is the suggestion from a report by Elle Decor and The Joy of Plants based on underpinning research from NASA and the American College of Allergy, Asthma, and Immunology published in *The Plantsman* horticultural journal.
24 Wolverton et al. (1989).
25 Holgate (2017).
26 Jones et al. (2019).
27 Facer-Childs et al. (2019).
28 Molteni (2017). www.wired.com/story/nobel-medicine-circadian-clocks/.
29 Agnew et al. (1966); Branstetter et al. (2012); Rattenborg et al. (1999); Tamaki et al. (2016).
30 'Best bedroom colors for sleep' (2020), 4 February, https://oursleepguide.com/best-bedroom-colors-for-sleep/; Costa et al. (2018).
31 *Guardian*, 4 September 2018, www.theguardian.com/lifeandstyle/2018/sep/04/shattered-legacy-of-a-reality-tv-experiment-in-extreme-sleep-deprivation; though, in fact, it later emerged that the contestants were allowed to take periodic forty-five-minute naps.
32 Kyle et al. (2010).
33 Field et al. (2008); Mindell et al. (2009).
34 Johnson's *Science of the senses* report (2015), www.johnsonsbaby.co.uk/healthcare-professionals/science-senses.
35 American Academy of Pediatrics, School start times for adolescents, Policy Statement, August 2014. See also www.startschoollater.net/success-stories.html; National Sleep Foundation (2006); Walker (2018).
36 Kaplan et al. (2019).

37 Holmes et al. (2002). See also Burns et al. (2002).
38 Crowley (2011); Hardy et al. (1995).
39 Fismer and Pilkington (2012).
40 Harada et al. (2018); Spence (2003).
41 Stumbrys et al. (2012); *Wired*, 31 March 2014, www.wired.co.uk/news/archive/2014-03/31/touch-taste-and-smell-technology.
42 Lovato and Lack (2016).
43 https://today.yougov.com/topics/lifestyle/articles-reports/2011/05/05/brother-do-you-have-time; Badia et al. (1990); *AdWeek*, 6 March 2014, www.adweek.com/adfreak/wake-and-smell-bacon-free-alarm-gadget-oscar-mayer-156123; Carskadon and Herz (2004); *Guardian*, 6 March 2014, www.theguardian.com/technology/2014/mar/06/wake-up-and-smell-the-bacon-scented-iphone-alarm-clock; *Intelligencer*, 29 November 2018, http://nymag.com/intelligencer/2018/11/iphone-bedtime-features-has-hidden-alarm-sounds.html.
44 Smith et al. (2006).
45 Broughton (1968); Jewett et al. (1999); Trotti (2017).
46 Fukuda and Aoyama (2017); Hilditch et al. (2016); *Vice*, 21 December 2015, www.vice.com/en_us/article/3dan5v/caffeinated-toothpaste-is-the-closest-youll-ever-get-to-mainlining-coffee.
47 Anderson et al. (2012); Government of India, Ministry of Civil Aviation, *Report on Accident to Air India Express Boeing 737-800 Aircraft VT-AXV on 22nd May 2010 at Mangalore*, www.skybrary.aero/bookshelf/books/1680.pdf; Schaefer et al. (2012); Tassi and Muzet (2000); Walker (2018).
48 McFarlane et al. (2020).
49 Gabel et al. (2013); Wright and Czeisler (2002).
50 Lamote de Grignon Pérez et al. (2019); Morosini (2019); Pinker (2018); Przybylski (2019).

## 5. Commuting

1. Redelmeier and Tibshirani (1997).
2. Colvile (2017).
3. Novaco et al. (1990).
4. www.volpe.dot.gov/news/how-much-time-do-americans-spend-behind-wheel.
5. Aikman (1951).
6. Harley-Davidson even went so far as to try to protect the 'potato-potato-potato' sound of their motorbike engine, but without success: Michael B. Sapherstein, The trademark registrability of the Harley-Davidson roar: A multimedia analysis, http://bciptf.org/wp-content/uploads/2011/07/48-THE-TRADEMARK-REGISTRABILITY-OF-THE-HARLEY.pdf.
7. *Sunday Times*, 12 June 2016. See also *Washington Post*, 21 January 2015.
8. Hellier et al. (2011).
9. Horswill and Plooy (2008).
10. Menzel et al. (2008).
11. *The Times*, 7 May 2018, 6.
12. Montignies et al. (2010).
13. BBC News, 14 January 2005, http://news.bbc.co.uk/go/pr/fr/-/2/hi/uk_news/wales/4174543.stm.
14. Sheldon and Arens (1932), pp. 100–101.
15. Guéguen et al. (2012); Hanss et al. (2012). See also Feldstein and Peli (2020).
16. Brodsky (2002); North and Hargreaves (1999).
17. Beh and Hirst (1999).
18. Ramsey and Simmons (1993).
19. *The Times*, 7 March 2018, 17.
20. Redelmeier and Tibshirani (1997).
21. Spence (2014).
22. Spence and Read (2003).

23 *New York Times*, 27 July 2009, www.nytimes.com/2009/07/28/technology/28texting.html; Driver distraction in commercial vehicle operations, Technical Report No. FMCSA-RRR-09-042, Federal Motor Carrier Safety Administration, US Department of Transportation, Washington, DC, 2009.
24 Ho and Spence (2008).
25 Ho and Spence (2009).
26 Obst et al. (2011).
27 Ashley (2001); Graham-Rowe (2001); Mitler et al. (1988); Sagberg (1999).
28 Oyer and Hardick (1963).
29 McKeown and Isherwood (2007).
30 Ho and Spence (2008).
31 *The Times*, 19 January 2018, 35.
32 Ho and Spence (2008).
33 Senders et al. (1967); Sivak (1996).
34 Cackowski and Nasar (2003).
35 Parsons et al. (1998).
36 This was certainly the intuition of Gibson and Crooks (1938).
37 Bijsterveld et al. (2014).
38 *De re aedificatoria* (1485), quoted in Lay (1992), p. 314.
39 *New York Times*, 5 July 2002, F1, www.nytimes.com/2002/07/05/travel/driving-just-drive-said-the-road-and-the-car-responded.html.
40 Gubbels (1938), p. 7.
41 Ury et al. (1972).
42 *New Atlas*, 26 January 2005, https://newatlas.com/go/3643/.
43 2014 Mercedes-Benz S-Class interior is 'the essence of luxury', https://emercedesbenz.com/autos/mercedes-benz/s-class/2014-mercedes-benz-s-class-interior-is-the-essence-of-luxury/.
44 Spence et al. (2017).
45 Forster and Spence (2018).
46 Ho and Spence (2005); Warm et al. (1991).

47 Fruhata et al. (2013); see also *Wall Street Journal*, 6 May 1996, B1, B5.
48 Fumento (1998); James and Nahl (2000); *2011 AAMI Crash Index*, www.yumpu.com/en/document/view/51279966/2011-aami-crash-index.
49 Mustafa et al. (2016).
50 Schiffman and Siebert (1991).
51 Ho and Spence (2013).
52 Evans and Graham (1991); Peltzman (1975); Wilde (1982).
53 Spence (2012a).
54 Whalen et al. (2004).
55 Treisman (1977).
56 Körber et al. (2015).
57 *The Times*, 24 January 2018, 26.
58 Deloitte, *Driving connectivity. Global automotive consumer study: Future of automotive technologies*, https://www2.deloitte.com/content/dam/Deloitte/uk/Documents/manufacturing/deloitte-uk-driving-connectivity.pdf, March 2017.
59 Where to, sir? *The Investor*, 95 (2017), 7–10.

## 6. Workplace

1 *Daily Mail*, 30 March 2006.
2 Hewlett and Luce (2006).
3 *The Economist*, 22 December 2018, www.economist.com/finance-and-economics/2018/12/22/why-americans-and-britons-work-such-long-hours; Pencavel (2014); *Wall Street Journal*, 29 June 2018.
4 *The Australian*, 7 December 2017; *Business Journal*, 11 June 2013, www.gallup.com/businessjournal/162953/tackleemployees-stagnating-engagement.aspx; Pencavel (2014).
5 *The Australian*, 7 December 2017; Béjean and Sultan-Taïeb (2005); *The Times*, 19 June 2017, 3.

6. Field et al. (1996).
7. Rosenthal (2019); Terman (1989).
8. Dolan (2004); Hirano (1996).
9. *The Economist*, 28 September 2019, www.economist.com/business/2019/09/28/redesigning-the-corporate-office; ibid., www.economist.com/leaders/2019/09/28/even-if-wework-is-in-trouble-the-office-is-still-being-reinvented; Haslam and Knight (2010); Knight and Haslam (2010).
10. Burge et al. (1987); Wargocki et al. (2000).
11. *Independent*, 14 May 2018, www.independent.co.uk/news/long_reads/sick-building-syndrome-treatment-finland-health-mould-nocebo-a8323736.html; Wargocki et al. (1999).
12. Baron (1994).
13. www.fellowes.com/gb/en/resources/fellowes-introduces/work-colleague-of-the-future.aspx; The work colleague of the future: A report on the long-term health of office workers, July 2019, https://assets.fellowes.com/skins/fellowes/responsive/gb/en/resources/work-colleague-of-the-future/download/WCOF_Report_EU.pdf. See also https://us.directlyapply.com/future-of-the-remote-worker.
14. Chang and Kajackaite (2019); Kingma and van Marken Lichtenbelt (2015).
15. Spence (2020d).
16. Küller et al. (2006).
17. Kozusznik et al. (2019); Pasut et al. (2015).
18. Pencavel (2014).
19. Kaida et al. (2006); Souman et al. (2017).
20. Fox and Embrey (1972); Oldham et al. (1995); Ross (1966); *Time*, 10 December 1984, 110–12.
21. Spence (2002, 2003).
22. Gabel et al. (2013); Lehrl et al. (2007).
23. Kwallek and Lewis (1990); Mikellides (1990).
24. *New York Times*, 5 February 2009, www.nytimes.com/2009/02/06/science/06color.html; Steele (2014).

25 *Wired*, 13 February 2019, www.wired.co.uk/article/how-workplace-design-can-foster-creativity.
26 Mehta, Zhu and Cheema (2012).
27 Einöther and Martens (2013); *Guardian*, 5 January 2014, www.theguardian.com/money/shortcuts/2014/jan/05/coffice-future-of-work; Madzharov et al. (2018); Unnava et al. (2018).
28 BBC News, 11 January 2017, www.bbc.com/capital/story/20170105-open-offices-are-damaging-our-memories.
29 Bernstein and Turban (2018); Otterbring et al. (2018).
30 De Croon et al. (2005), p. 128; *The Times*, 10 October 2017, 6–7.
31 *Guardian*, 16 October 2015, www.theguardian.com/higher-education-network/2015/oct/16/the-open-plan-university-noisy-nightmare-or-buzzing-ideas-hub.
32 Yildirim et al. (2007).
33 Levitin (2015). See also *Forbes*, 21 June 2016, www.forbes.com/sites/davidburkus/2016/06/21/why-your-open-office-workspace-doesnt-work/#188f073a435f; Evans and Johnson (2000); *The Times*, 10 October, 6–7.
34 Hongisto et al. (2017).
35 Haga et al. (2016).
36 Leather et al. (1998); Mitchell and Popham (2008).
37 Bringslimark et al. (2011); Kweon et al. (2008).
38 Nieuwenhuis et al. (2014).
39 Krieger (1973), p. 453. See also Wohlwill (1983).
40 Qin et al. (2014), though see Cummings and Waring (2020) for recent evidence questioning the practical benefit of office plants when it comes to removing VOCs from the air.
41 Guieysse et al. (2008); Wood et al. (2006).
42 *Raconteur*, 24 April 2019, 8, on the benefits of biophilic office design.
43 Berman et al. (2008); Berto (2005).
44 Lee et al. (2015).
45 De Kort et al. (2006).
46 Kahn et al. (2008).

47 Annerstedt et al. (2013).
48 Spence (2002).
49 Gillis and Gatersleben (2015); Spence (2002).
50 Spence (2016).
51 Quote from *Forbes*, 2 July 2015, www.forbes.com/sites/davidburkus/2015/07/02/the-real-reason-google-serves-all-that-free-food/#7e426b603e3b.
52 Balachandra (2013); Kniffin et al. (2015); Woolley and Fishbach (2017).
53 *New York Times*, 2 July 2012, D3, www.nytimes.com/2012/07/04/dining/secretary-of-state-transforms-the-diplomatic-menu.html?_r=0.

## 7. Shopping

1 *Marketing Week*, 31 October 2013, www.marketingweek.com/2013/10/30/sensory-marketing-could-it-be-worth-100m-to-brands/; Hilton (2015).
2 *Financial Times*, 4 June 2013, 1.
3 Samuel (2010).
4 Renvoisé and Morin (2007); Kühn et al. (2016).
5 Aiello et al. (2019) for a recent analysis of 1.6 billion fidelity card transactions; *Venture Beat*, 11 February 2019, https://venturebeat.com/2019/02/11/second-measure-raises-20-million-to-analyze-companies-sales-and-growth-rates/.
6 *Independent*, 16 August 2011, www.independent.co.uk/news/media/advertising/the-smell-of-commerce-how-companies-use-scents-to-sell-their-products-2338142.html; *Time*, 20 July 2011, http://business.time.com/2011/07/20/nyc-grocery-store-pipes-in-artificial-food-smells/.
7 *Wall Street Journal*, 20 May 2014, www.wsj.com/articles/SB10001424052702303468704579573953132979382; Spence (2015).
8 Leenders et al. (2019).
9 Spence et al. (2017).

10 *Independent*, 16 August 2011, www.independent.co.uk/news/media/advertising/the-smell-of-commerce-how-companies-use-scents-to-sell-their-products-2338142.html.
11 Ayabe-Kanamura et al. (1998).
12 Spence and Carvalho (2020).
13 *NACS Magazine*, 8–9 August 2009, www.scentandrea.com/MakesScents.pdf.
14 *The Atlantic*, 26 July 2012, www.theatlantic.com/technology/archive/2012/07/the-future-of-advertising-will-be-squirted-into-your-nostrils-as-you-sit-on-a-bus/260283/.
15 Knoeferle et al. (2016); Spence (2019a).
16 *AdAge*, 6 December 2006, http://adage.com/article/news/milk-board-forced-remove-outdoor-scent-strip-ads/113643/.
17 *Independent*, 14 November 2002, www.independent.co.uk/news/media/whiff-of-almond-falls-victim-to-terror-alert-133417.html; Lim (2014), p. 84.
18 *CityLab*, 9 February 2012, www.citylab.com/design/2012/02/inside-smellvertising-scented-advertising-tactic-coming-bus-stop-near-you/1181/; McCain creates the world's first potato scented taxi – Offering free hot jacket potatoes that are cooked on board in five minutes! Press Release, 9 November, 2013; Metcalfe (2012).
19 Castiello et al. (2006).
20 *Businessweek*, 17 October 2013, www.businessweek.com/articles/2013-10-17/chipotles-music-playlists-created-by-chris-golub-of-studio-orca; Milliman (1982, 1986); see also Mathiesen et al. (2020).
21 Lanza (2004).
22 Knoeferle et al. (2012).
23 North, Hargreaves and McKendrick (1997).
24 Spence et al. (2019b); Zellner et al. (2017).
25 Karremans et al. (2006).
26 *Economist 1843 Magazine*, April/May 2019, www.1843magazine.com/design/brand-illusions/why-stars-make-your-water-sparkle; Spence (2012b).

27 Kotler (1974); Lindstrom (2005).
28 *AdWeek*, 5 March 2012, www.adweek.com/brand-marketing/something-air-138683/.
29 *Wall Street Journal*, 24 November 2000; www.springwise.com/summer-jeans-embedded-aroma-fruit/.
30 Minsky et al. (2018).
31 Ayabe-Kanamura et al. (1998); Trivedi (2006).
32 *AdWeek*, 5 March 2012, www.adweek.com/brand-marketing/something-air-138683/.
33 Preliminary results of olfaction Nike study, note dated 16 November 1990, distributed by the Smell and Taste Treatment and Research Foundation Ltd, Chicago; *Marketing News*, 25, 4 February 1991, 1–2; though see *Chicago Tribune*, 19 January 2014, www.chicagotribune.com/lifestyles/health/ct-met-sensa-weight-loss-hirsch-20140119-story.html.
34 Knasko (1989); *Wall Street Journal*, 9 January 1990, B5.
35 *USA Today*, 1 September 2006; Trivedi (2006); *Independent*, 16 August 2011, www.independent.co.uk/news/media/advertising/the-smell-of-commerce-how-companies-use-scents-to-sell-their-products-2338142.html.
36 *New York Times*, 26 June 2005, www.nytimes.com/2005/06/26/fashion/sundaystyles/shivering-for-luxury.html; Park and Hadi (2020); quote appears in Tanizaki (2001), p. 10.
37 Martin (2012).
38 Peck and Shu (2009).
39 Ellison and White (2000); Spence and Gallace (2011).
40 Gallace and Spence (2014), Chapter 11.
41 Does it make sense? *Contact: Royal Mail's Magazine for Marketers*, Sensory marketing special edition, November 2007, 39; Solomon (2002).
42 *Forbes*, 14 June 2012, www.forbes.com/sites/carminegallo/2012/06/14/why-the-new-macbook-pro-is-tilted-70-degrees-in-an-apple-store/#784de2f65a98.

43 Hultén (2012).
44 Piqueras-Fiszman and Spence (2012).
45 Argo et al. (2006).
46 Underhill (1999), p. 162.
47 *Newsweek*, 28 November 2018, www.newsweek.com/mcdonalds-touchscreen-machines-tested-have-fecal-matter-investigation-finds-1234954.
48 de Wijk et al. (2018); Helmefalk and Hultén (2017).
49 Roschk et al. (2017); Schreuder et al. (2016).
50 Mattila and Wirtz (2001).
51 Morrin and Chebat (2005).
52 Homburg et al. (2012).
53 Malhotra (1984); Spence et al. (2014b).
54 Quoted in *Mail Online*, 23 May 2014, www.dailymail.co.uk/femail/article-2637492/Lights-sound-clothes-Abercrombie-Fitch-tones-nightclub-themed-stores-bid-win-disinterested-teens.html.
55 See Spence et al. (2019b) for a review.
56 Dunn (2007).
57 Malhotra (1984); *Canvas 8*, 18 January 2013, www.canvas8.com/public/2013/01/18/no-noise-selfridges.html.
58 Spence (2019b); *The Drum*, 18 May 2017, www.thedrum.com/news/2017/05/18/guinness-tantalises-tesco-shoppers-with-vr-tasting-experience; *VR Focus*, 20 May 2017, www.vrfocus.com/2017/05/vr-in-the-supermarket-with-guinness-vr-tasting-experience/.
59 Petit et al. (2019).
60 Kampfer et al. (2017).
61 Gallace and Spence (2014).
62 *RFID Journal*, 14 September 2017, www.rfidjournal.com/articles/pdf?16605; *ShopifyPlus*, 27 February 2019, www.shopify.com/enterprise/ecommerce-returns.
63 Jütte (2005).
64 Spence et al. (2017).
65 Spence (2020a,b); Spence et al. (2020).

## 8. Healthcare

1. EurekAlert, 20 December 2014, www.eurekalert.org/pub_releases/2014-12/bmj-woy121014.php; Ullmann et al. (2008).
2. Lies and Zhang (2015).
3. Allen and Blascovich (1994).
4. Shippert (2005).
5. Fancourt et al. (2016).
6. Hawksworth et al. (1997).
7. Gatti and da Silva (2007).
8. Kotler (1974).
9. *Forbes*, 18 June 2018, www.forbes.com/sites/brucejapsen/2018/06/18/more-doctor-pay-tied-to-patient-satisfaction-and-outcomes/#567codb1504a.
10. *Telegraph*, 22 June 2019, www.telegraph.co.uk/health-fitness/body/looks-like-hotel-best-hospital-world-opening-doors-london/.
11. Richter and Muhlestein (2017). See also https://blog.experientia.com/reinventing-cancer-surgery-by-designing-a-better-hospital-experience/.
12. Trzeciak et al. (2016).
13. Richter and Muhlestein (2017).
14. Ottoson and Grahn (2005); Ulrich (1999).
15. Franklin (2012).
16. Antonovsky (1979); Zhang et al. (2019); Nightingale (1860); Ulrich (1991).
17. Spence and Keller (2019).
18. Spence (2017).
19. Ziegler (2015).
20. *Telegraph*, 12 January 2019, www.telegraph.co.uk/news/2019/01/12/giving-elderlyhospital-patients-one-extra-meal-day-cuts-deaths/.
21. Campos et al. (2019).
22. Spence (2017).

23 Palmer (1978).
24 *Smithsonian Magazine*, 3 May 2018, www.smithsonianmag.com/smithsonian-institution/could-our-housewares-keep-us-healthier-180968950/; *Wired*, 3 October 2015, www.wired.co.uk/magazine/archive/2015/11/play/lizzie-ostrom-smell.
25 Dijkstra et al. (2008).
26 Lankston et al. (2010).
27 Harper et al. (2015).
28 Tse et al. (2002); Staricoff and Loppert (2003).
29 Nightingale (1860); see also *Telegraph*, 22 June 2019, www.telegraph.co.uk/health-fitness/body/looks-like-hotel-best-hospital-world-opening-doors-london/.
30 Pancoast (1877); Babbitt (1896).
31 Dalke et al. (2006).
32 *Telegraph*, 22 June 2019, www.telegraph.co.uk/health-fitness/body/looks-like-hotel-best-hospital-world-opening-doors-london/; www.philips.co.uk/healthcare/consulting/experience-solutions/ambient-experience; www.itsnicethat.com/news/g-f-smith-most-relaxing-colour-survey-miscellaneous-100419.
33 Ramachandran and Blakeslee (1998); Senkowski et al. (2014).
34 Moseley et al. (2008a).
35 Moseley et al. (2008c).
36 Moseley et al. (2008b).
37 Barnsley et al. (2011); Mancini et al. (2011); Wittkopf et al. (2018).
38 This quote appears in Katz (2014).
39 Rice (2003).
40 Darbyshire (2016); Darbyshire and Young (2013).
41 Berglund et al. (1999).
42 Yoder et al. (2012).
43 *Telegraph*, 30 March 2016, www.telegraph.co.uk/news/science/science-news/12207648/critically-ill-patients-disturbed-every-six-minutes-at-night-in/.

44 *Telegraph*, 15 April 2016, www.telegraph.co.uk/science/2016/04/15/cambridge-professor-reduced-to-tears-by-noisy-hospital-before-de/.
45 Rybkin (2017); Siverdeen et al. (2008).
46 Carlin et al. (1962).
47 Stanton et al. (2017).
48 Diette et al. (2003); Villemure et al. (2003).
49 This is graphically captured in the opening pages of Dan Ariely's *Predictably irrational* (2008).
50 *Wired*, 2 November 2018, www.wired.com/story/opioids-havent-solved-chronic-pain-maybe-virtual-reality-can/; Li et al. (2011).
51 *Guardian*, 25 January 2017, www.theguardian.com/science/2017/jan/25/how-doctors-measure-pain/.
52 Spence and Keller (2019).
53 Conrad et al. (2007).
54 Graff et al. (2019); Spence and Keller (2019) for a review.
55 See Spence and Keller (2019) for a review.
56 Moss et al. (2007).
57 *Independent*, 18 April 2013, www.independent.co.uk/arts-entertainment/art/news/from-roxy-music-to-the-cure-brian-eno-composes-soundscapes-to-treat-hospital-patients-8577179.html.
58 Field (2001); *The Conversation*, 24 May 2016, https://theconversation.com/touch-creates-a-healing-bond-in-health-care-59637.
59 Ellingsen et al. (2016).
60 Gallace and Spence (2014).
61 Crossman (2017).
62 Hamilton (1966).
63 Prescott and Wilkie (2007).
64 Blass and Shah (1995).
65 Holmes et al. (2002).
66 Lehrner et al. (2000).
67 Fenko and Loock (2014).

68 Hulsegge and Verheul (1987).
69 See http://go.ted.com/bUcH for why multisensory palliative care is such a good idea.

## 9. Exercise and Sport

1 Hillman et al. (2008).
2 Mead et al. (2009); Chekroud et al. (2018).
3 Craig et al. (2009).
4 NHS Digital, Health Survey for England 2018, https://digital.nhs.uk/data-and-information/publications/statistical/health-survey-for-england/2018.
5 *Guardian*, 8 May 2017, www.theguardian.com/lifeandstyle/shortcuts/2017/may/08/the-budget-gym-boom-how-low-cost-clubs-are-driving-up-membership.
6 *CityLab*, 2 January 2018, www.bloomberg.com/news/articles/2018-01-02/the-geography-of-the-urban-fitness-boom.
7 Though, as we will see later, it may be more the fast music than the loud music that is doing the work; Kreutz et al. (2018).
8 Bodin and Hartig (2003).
9 Thompson Coon et al. (2011).
10 Deloitte, *Health of the nation* (2006), cited in Thompson Coon et al. (2011).
11 RSPB, Natural fit. Can green space and biodiversity increase levels of physical activity? (2004), http://ww2.rspb.org.uk/Images/natural_fit_full_version_tcm9-133055.pdf.
12 *Mail Online*, 13 May 2018, www.dailymail.co.uk/news/article-5723627/David-Lloyd-launches-personal-trainers-TV-screens-backs.html.
13 *The Times*, 12 May 2018, www.thetimes.co.uk/article/the-latest-fitness-trend-the-cavewoman-workout-38jgqjsfg.

14  Plante et al. (2006).
15  Williams (2017), pp. 176–8.
16  Morgan et al. (1988).
17  Raudenbush et al. (2002).
18  Barwood et al. (2009); North et al. (1998).
19  Karageorghis and Terry (1997).
20  Bigliassi et al. (2019); see also Suwabe et al. (2020).
21  Beach and Nie (2014); *Chicago Tribune*, 17 February 2014, www.chicagotribune.com/lifestyles/health/chi-gym-loud-music-20150218-story.html.
22  Waterhouse et al. (2010).
23  Patania et al. (2020).
24  Edworthy and Waring (2006).
25  Terry et al. (2012).
26  Fritz et al. (2013).
27  North and Hargreaves (2000); Priest et al. (2004).
28  Schaffert et al. (2011).
29  *Guardian*, 17 January 2018, www.theguardian.com/sport/2018/jan/17/noise-over-grunting-cranks-up-once-again-after-crowd-mocks-aryna-sabalenka.
30  *Mail Online*, 7 June 2018, www.dailymail.co.uk/news/article-5818615/Greg-Rusedski-says-women-tennis-players-louder-747-aeroplane.html.
31  Cañal-Bruland et al. (2018).
32  Sinnett and Kingstone (2010).
33  Müller et al. (2019).
34  Quoted in Sinnett and Kingstone (2010).
35  BBC News, 17 May 2009, http://news.bbc.co.uk/sport1/hi/tennis/7907707.stm.
36  Camponogara et al. (2017); Sors et al. (2017).
37  Unkelbach and Memmert (2010).
38  Balmer et al. (2005).

39 Raudenbush et al. (2001); Raudenbush et al. (2002).
40 Romine et al. (1999).
41 Brick et al. (2018); www.bbc.co.uk/sport/athletics/50025543.
42 Chambers et al. (2009).
43 Researchers at Oxford have even developed a brand-new sports drink that improves the body's ability to transport oxygen direct to the muscles: *The Times*, 5 May 2020, www.thetimes.co.uk/article/is-an-energy-drink-that-supplies-oxygen-to-the-muscles-the-ultimate-performance-booster-cmhm6stgq.
44 Carter et al. (2004).
45 Ibid.
46 Ataide-Silva et al. (2014).
47 Hollingworth (1939); Scholey et al. (2009); though see Walker et al. (2016).
48 *Guardian*, 17 May 2012, www.theguardian.com/football/2012/may/17/wayne-rooney-visualisation-preparation.
49 Wrisberg and Anshel (1989).
50 Frank and Gilovich (1988).
51 Huang et al. (2011).
52 Hill and Barton (2005).
53 Attrill et al. (2008).
54 Hagemann et al. (2008).
55 Barton and Hill (2005); Rowe et al. (2005).
56 Elliot et al. (2007).
57 Hill and Barton (2005).
58 Changizi et al. (2006).
59 Phalen (1910), cited in http://history.amedd.army.mil/booksdocs/spanam/gillet3/bib.html.
60 Adam and Galinsky (2012).
61 *Telegraph*, 31 May 2014, www.telegraph.co.uk/news/science/science-news/10866021/Wear-a-Superman-t-shirt-to-boost-exam-success.html.

## 10. Dating

1 Groyecka et al. (2017).
2 Dutton and Aron (1974).
3 Meston and Frohlich (2003).
4 Cohen et al. (1989).
5 Marin et al. (2017).
6 May and Hamilton (1980).
7 Hove and Risen (2009).
8 Byers et al. (2010).
9 Hugill et al. (2010); McCarty et al. (2017); Neave et al. (2011).
10 Grammer et al. (2004).
11 Roberts et al. (2004).
12 Miller et al. (2007).
13 Rhodes (2006).
14 Jones et al. (2018); Mueser et al. (1984).
15 Abel and Kruger (2010).
16 Liu et al. (2015).
17 Kampe et al. (2001).
18 Tifferet et al. (2012).
19 Miller (2000).
20 Darwin (1871).
21 Charlton et al. (2012).
22 Watkins (2017).
23 Havlíček et al. (2008); Herz and Cahill (1997); Buss (1989).
24 Nettle and Pollet (2008).
25 Baker (1888); Manning and Fink (2008).
26 Manning et al. (1998).
27 Geschwind and Galaburda (1985).
28 Havlíček et al. (2006); Kuukasjärvi et al. (2004).
29 Lobmaier et al. (2018).
30 Sorokowska et al. (2012).

31 Olsson et al. (2014).
32 Mitro et al. (2012).
33 Roberts et al. (2011).
34 Winternitz et al. (2017).
35 According to Herz and Cahill (1997), more than $5 billion is spent annually on fragrance.
36 Lenochová et al. (2012); Milinski and Wedekind (2001).
37 *Guardian*, 24 March 2006, www.theguardian.com/education/2006/mar/24/schools.uk3.
38 *New Zealand Herald*, 19 February 2007, www.nzherald.co.nz/nz/news/article.cfm?c_id=1&objectid=10424667.
39 Demattè et al. (2007).
40 *Perfumer and Flavorist*, 1 April 2016, www.perfumerflavorist.com/fragrance/trends/A-Taste-of-Gourmand-Trends-374299261.html.
41 *The Economist*, 14 February 2008, www.economist.com/news/2008/02/14/food-of-love.
42 McGlone et al. (2013).
43 *Maxim*, March 2007, 132–3.
44 Li et al. (2007).
45 Griskevicius and Kenrick (2013), p. 379.
46 Elliot and Niesta (2008); Guéguen (2012).
47 Beall and Tracy (2013); Elliot and Pazda (2012).
48 Guéguen and Jacob (2014).
49 Stillman and Hensley (1980); Jacob et al. (2012).
50 Guéguen and Jacob (2011).
51 Jones and Kramer (2016).
52 Lin (2014); though see Pollet et al. (2018).
53 Greenfield (2005).
54 E.g. Lynn et al. (2016); Peperkoorn et al. (2016).
55 *Slate*, 24 July 2013, www.slate.com/articles/health_and_science/science/2013/07/statistics_and_psychology_multiple_comparisons_give_spurious_results.html.
56 Lewis et al. (2017).

57 Lewis et al. (2015).
58 Whitcome et al. (2007).
59 Tobin (2014). Online dating services, http://yougov.co.uk/news/2014/02/13/seven-ten-online-dating-virgins-willing-try-findin/.
60 Willis and Todorov (2006).
61 White et al. (2017).
62 *Guardian*, 11 February 2011, www.theguardian.com/lifeandstyle/wordofmouth/2011/feb/11/aphrodisiacs-food-of-love.
63 Otterbring (2018).
64 Gladue and Delaney (1990).
65 Jones et al. (2003).
66 Chen et al. (2014).
67 See note 62 above.
68 Apicella et al. (2007).
69 Feinberg et al. (2008).
70 Pavela Banai (2017).
71 Ratcliffe et al. (2016).
72 McGuire et al. (2018).
73 Groyecka et al. (2017).
74 Miller (1998).
75 Roche (2019); *Independent*, 10 August 2017, www.independent.co.uk/life-style/11-scientific-ways-to-make-yourself-look-and-feel-more-attractive-a7886021.html.

## 11. Coming to Our Senses

1 Ackerman (2000); Rosenblum (2010).
2 Cabanac (1979); Pfaffmann (1960).
3 Merabet et al. (2004); Motluck (2007).
4 *Science Alert*, 20 November 2019, www.sciencealert.com/dopamine-fasting-is-silicon-valley-s-latest-trend-here-s-what-an-expert-has-to-say.

5 *New York Times*, 7 November 2019, www.nytimes.com/2019/11/07/style/dopamine-fasting.html; *The Times*, 19 November 2019, 27.
6 Kranowitz (1998); Longman (2019). *New York Times*, 1 November 2019, www.nytimes.com/2019/11/01/sports/football/eagles-sensory-disorder-autism.html.
7 Colvile (2017); see also https://www.nielsen.com/us/en/insights/article/2010/three-screen-report-q409/.
8 Milgram (1970), p. 1462; see also Blass (2004).
9 Barr (1970); Diaconu et al. (2011).
10 UN-Habitat, *State of the world's cities 2010/2011: Bridging the urban divide*, https://sustainabledevelopment.un.org/content/documents/11143016_alt.pdf.
11 Currently, somewhere around 55 per cent of the population live in urban areas, up from 30 per cent in 1950, see www.un.org/development/desa/en/news/population/2018-revision-of-world-urbanization-prospects.html.
12 Guieysse et al. (2008); Ott and Roberts (1998).
13 *New Yorker*, 13 May 2019, www.newyorker.com/magazine/2019/05/13/is-noise-pollution-the-next-big-public-health-crisis; Velux YouGov Report, 14 May 2018, https://press.velux.com/download/542967/theindoorgenerationsurvey14may2018-2.pdf.
14 Walker (2018).
15 National Trust press release, 27 February 2020, www.nationaltrust.org.uk/press-release/national-trust-launches-year-of-action-to-tackle-nature-deficiency-; Williams (2017).
16 Just et al. (2019).
17 Spence (2020d).
18 Malhotra (1984).
19 *Financial Times*, 4 June 2013, www.ft.com/content/3ac8eac6-cf93-11dc-854a-0000779fd2ac; *New Yorker*, 26 October 2012, www.newyorker.com/magazine/2015/11/02/accounting-for-taste.
20 Bremner et al. (2012); Calvert et al. (2004); Stein (2012).

21 Spence (2020b).
22 Kabat-Zinn (2005); Spence (2002).
23 Achrol and Kotler (2011), p. 37.
24 *The Conversation*, 2 August 2018, http://theconversation.com/the-coded-images-that-let-advertisers-target-all-our-senses-at-once-98676; *The Wired World in 2013*, November 2012, 104–7.
25 Balken (1997); Haverkamp (2014); Marks (1978); Zilczer (1987).
26 Spence (2002).
27 Field (2001); Harlow and Zimmerman (1959); *The Times*, 17 February 2020, www.thetimes.co.uk/article/how-to-greet-in-2020-what-is-and-what-isnt-appropriate-qq7jqxrrv.
28 Denworth (2015); Sekuler and Blake (1987).
29 Cain (2012); Zuckerman (1979).
30 Longman (2019); Lupton and Lipps (2018).
31 Hutmacher (2019); Le Breton (2017); Levin (1993); McGann (2017).
32 Keller (2008); Smith (2007).
33 *New Yorker*, 23 March 2020, www.newyorker.com/news/our-columnists/how-loneliness-from-coronavirus-isolation-takes-its-own-toll.
34 Block and Kuchenbecker (2018); Cute Circuit, https://cutecircuit.com/media/the-hug-shirt/; *Time*, Best inventions of 2006, http://content.time.com/time/specials/packages/article/0,28804,1939342_1939424_1939709,00.html; *The Times*, 12 June 2018, www.thetimes.co.uk/article/strong-and-non-clingy-robots-give-the-best-hugs-study-reveals-huggiebot-pdx566xk0; Geddes (2020).
35 Gallace and Spence (2014).
36 Kahn et al. (2009); Krieger (1973).
37 Spence et al. (2019a).
38 *Smithsonian Magazine*, 18 January 2017, www.smithsonianmag.com/innovation/artificial-sixth-sense-helps-humans-orient-themselves-world-180961822/; see also www.cyborgarts.com/.

39 Neil Harbisson, I listen to colour, TEDGlobal, June 2012, www.ted.com/talks/neil_harbisson_i_listen_to_color.html; Gafsou and Hildyard (2019).
40 Quoted in Bainbridge (2018).
41 *Mail Online*, 16 May 2017, www.dailymail.co.uk/news/article-4509940/Man-compass-implanted-chest.html.
42 Kurzweil (2005); O'Connell (2018).
43 *Wall Street Journal*, 27 March 2017, www.wsj.com/articles/elon-musk-launches-neuralink-to-connect-brains-with-computers-1490642652.
44 Spence (2020f).
45 Howes (2004); Howes (2014); Howes and Classen (2014); Schwartzman (2011).

# Bibliography

Abel, E. L. and Kruger, M. L. (2010). Smile intensity in photographs predicts longevity. *Psychological Science*, 21, 542–4

Achrol, R. S. and Kotler, P. (2011). Frontiers of the marketing paradigm in the third millennium. *Journal of the Academy of Marketing Science*, 40, 35–52

Ackerman, D. (2000). *A natural history of the senses.* London: Phoenix

Adam, D. (2018). *The genius within: Smart pills, brain hacks and adventures in intelligence.* London: Picador

Adam, H. and Galinsky, A. D. (2012). Enclothed cognition. *Journal of Experimental Social Psychology*, 48, 918–25

Agnew, H. W., Jr et al. (1966). The first night effect: An EEG study of sleep. *Psychophysiology*, 2, 263–6

Aiello, L. M. et al. (2019). Large-scale and high-resolution analysis of food purchases and health outcomes. *EPJ Data Science*, 8, 14

Aikman, L. (1951). Perfume, the business of illusion. *National Geographic*, 99, 531–50

Allen, K. and Blascovich, J. (1994). Effects of music on cardiovascular reactivity among surgeons. *Journal of the American Medical Association*, 272, 882–4

Alter, A. (2013). *Drunk tank pink: And other unexpected forces that shape how we think, feel, and behave.* New York: Penguin

Alvarsson, J. J. et al. (2010). Nature sounds beneficial: Stress recovery during exposure to nature sound and environmental noise. *International Journal of Environmental Research and Public Health*, 7, 1036–46

Ambrose, G. et al. (2020). Is gardening associated with greater happiness of urban residents? A multiactivity, dynamic assessment in the Twin-Cities region, USA. *Landscape and Urban Planning*, 198, 103776

Ames, B. N. (1989). Pesticides, risk, and applesauce. *Science*, 244, 755–7

Anderson, C. et al. (2012). Deterioration of neurobehavioral performance in resident physicians during repeated exposure to extended duration work shifts. *Sleep*, 35, 1137–46

Anderson, L. M. et al. (1983). Effects of sounds on preferences for outdoor settings. *Environment and Behavior*, 15, 539–66

Annerstedt, M. et al. (2013). Inducing physiological stress recovery with sounds of nature in a virtual reality forest – results from a pilot study. *Physiology and Behavior*, 118, 240–50

Antonovsky, A. (1979). *Health, stress and coping*. San Francisco: Jossey-Bass

Apicella, C. L. et al. (2007). Voice pitch predicts reproductive success in male hunter-gatherers. *Biology Letters*, 3, 682–4

Appleton, J. (1975). *The experience of landscape*. New York: John Wiley & Sons (repr. 1996)

Appleyard, D., Lynch, K. and Myer, J. R. (1965). *The view from the road*. Cambridge, MA: MIT Press.

Arbon, E. L. et al. (2015). Randomised clinical trial of the effects of prolonged release melatonin, temazepam and zolpidem on slow-wave activity during sleep in healthy people. *Journal of Psychopharmacology*, 29, 764–76

Argo, J. et al. (2006). Consumer contamination: How consumers react to products touched by others. *Journal of Marketing*, 70 (April), 81–94

Ariely, D. (2008). *Predictably irrational: The hidden forces that shape our decisions*. London: HarperCollins

Arzi, A. et al. (2012). Humans can learn new information during sleep. *Nature Neuroscience*, 15, 1460–65

Ashley, S. (2001). Driving the info highway. *Scientific American*, 285, 44–50

Ataide-Silva, T. et al. (2014). Can carbohydrate mouth rinse improve performance during exercise? A systematic review. *Nutrients*, 6, 1–10

Attfield, J. (1999). Bringing modernity home: Open plan in the British domestic interior. In I. Cieraad (ed.), *At home: An anthropology of domestic space*. New York: Syracuse University Press, pp. 73–82

Attrill, M. J. et al. (2008). Red shirt colour is associated with long-term team success in English football. *Journal of Sports Sciences*, 26, 577–82

Ayabe-Kanamura, S. et al. (1998). Differences in perception of everyday odors: A Japanese–German cross-cultural study. *Chemical Senses*, 23, 31–8

Babbitt, E. D. (1896). *The principles of light and color*. East Orange, NJ: Published by the author

Badia, P. et al. (1990). Responsiveness to olfactory stimuli presented in sleep. *Physiology and Behavior*, 48, 87–90

Bailly Dunne, C. and Sears, M. (1998). *Interior designing for all five senses*. New York: St. Martin's Press

Baird, J. C. et al. (1978). Room preference as a function of architectural features and user activities. *Journal of Applied Psychology*, 63, 719–27

Baker, F. (1888). Anthropological notes on the human hand. *American Anthropologist*, 1, 51–76

Balachandra, L. (2013). Should you eat while you negotiate? *Harvard Business Review*, 29 January, https://hbr.org/2013/01/should-you-eat-while-you-negot

Balken, D. B. (1997). *Arthur Dove: A retrospective*. Cambridge, MA: MIT Press

Balmer, N. J. et al. (2005). Do judges enhance home advantage in European championship boxing? *Journal of Sports Sciences*, 23, 409–16

Barnsley, N. et al. (2011). The rubber hand illusion increases histamine reactivity in the real arm. *Current Biology*, 21, R945–R946

Baron, R. A. (1994). The physical environment of work settings: Effects on task performance, interpersonal relations, and job satisfaction. In B. M. Staw and L. L. Cummings (eds.), *Research in organizational behaviour*, 16, pp. 1–46

——— (1997). The sweet smell of helping: Effects of pleasant ambient fragrance on prosocial behavior in shopping malls. *Personality and Social Psychology Bulletin*, 23, 498–505

Barr, J. (1970). *The assaults on our senses*. London: Methuen

Barton, R. A. and Hill, R. A. (2005). Sporting contests – seeing red? Putting sportswear in context – Reply. *Nature*, 437, E10–E11

Barwood, M. J. et al. (2009). A motivational music and video intervention improves high-intensity exercise performance. *Journal of Sports Science and Medicine*, 8, 435–42

Basner, M. et al. (2014). Auditory and non-auditory effects of noise on health. *The Lancet*, 383, 1325–32

Batra, R. et al. (eds.) (2016). *The psychology of design: Creating consumer appeal*. London: Routledge

Beach, E. F. and Nie, V. (2014). Noise levels in fitness classes are still too high: Evidence from 1997–1998 and 2009–2011. *Archives of Environmental and Occupational Health*, 69, 223–30

Beall, A. T. and Tracy, J. L. (2013). Women are more likely to wear red or pink at peak fertility. *Psychological Science*, 24, 1837–41

Beh, H. C. and Hirst, R. (1999). Performance on driving-related tasks during music. *Ergonomics*, 42, 1087–98

Béjean, S. and Sultan-Taïeb, H. (2005). Modeling the economic burden of diseases imputable to stress at work. *European Journal of Health Economics*, 6, 16–23

Bell, G. and Kaye, J. (2002). Designing technology for domestic spaces: A kitchen manifesto. *Gastronomica*, 2, 46–62

Bellak, L. (1975). *Overload: The new human condition*. New York: Human Sciences Press

Benfield, J. A. et al. (2010). Does anthropogenic noise in national parks impair memory? *Environment and Behavior*, 42, 693–706

Berglund, B. et al. (1999). *Guidelines for community noise*. Geneva: World Health Organization

Berman, M. G. et al. (2008). The cognitive benefits of interacting with nature. *Psychological Science*, 19, 1207–12

Bernstein, E. S. and Turban, S. (2018). The impact of the 'open' workspace on human collaboration. *Philosophical Transactions of the Royal Society B*, 373, 20170239

Berto, R. (2005). Exposure to restorative environments helps restore attentional capacity. *Journal of Environmental Psychology*, 25, 249–59

Bigliassi, M. et al. (2019). The way you make me feel: Psychological and cerebral responses to music during real-life physical activity. *Journal of Sport and Exercise*, 41, 211–17

Bijsterveld, K. et al. (2014). *Sound and safe: A history of listening behind the wheel*. Oxford: Oxford University Press

Blass, E. M. and Shah, A. (1995). Pain reducing properties of sucrose in human newborns. *Chemical Senses*, 20, 29–35

Blass, T. (2004). *The man who shocked the world: The life and legacy of Stanley Milgram*. New York: Basic Books

Block, A. E. and Kuchenbecker, K. J. (2018). Emotionally supporting humans through robot hugs. *HRI '18: Companion of the 2018 ACM/IEEE International Conference on Human–Robot Interaction, March 2018*, 293–4

Bodin, M. and Hartig, T. (2003). Does the outdoor environment matter for psychological restoration gained through running? *Psychology of Sport and Exercise*, 4, 141–53

Bowler, D. E. et al. (2010). A systematic review of evidence for the added benefits to health of exposure to natural environments. *BMC Public Health*, 10, 456

Branstetter, B. K. et al. (2012). Dolphins can maintain vigilant behavior through echolocation for 15 days without interruption or cognitive impairment. *PLOS One*, 7, e47478

Bratman, G. N. et al. (2015). Nature experience reduces rumination and subgenual prefrontal cortex activation. *Proceedings of the National Academy of Sciences of the USA*, 112, 8567–72

Bremner, A. et al. (eds.) (2012). *Multisensory development*. Oxford: Oxford University Press

Brick, N. et al. (2018). The effects of facial expression and relaxation cues on movement economy, physiological, and perceptual responses during running. *Psychology of Sport and Exercise*, 34, 20–28

Bringslimark, T. et al. (2011). Adaptation to windowlessness: Do office workers compensate for a lack of visual access to the outdoors? *Environment and Behavior*, 43, 469–87

Brodsky, W. (2002). The effects of music tempo on simulated driving performance and vehicular control. *Transportation Research Part F*, 4, 219–41

Broughton, R. J. (1968). Sleep disorders: Disorders of arousal? *Science*, 159, 1070–78

Bschaden, A. et al. (2020). The impact of lighting and table linen as ambient factors on meal intake and taste perception. *Food Quality and Preference*, 79, 103797

Buijze, G. A. et al. (2016). The effect of cold showering on health and work: A randomized controlled trial. *PLOS One*, 11, e0161749

Burge, S. et al. (1987). Sick building syndrome: A study of 4373 office workers. *Annals of Occupational Hygiene*, 31, 493–504

Burns, A. et al. (2002). Sensory stimulation in dementia: An effective option for managing behavioural problems. *British Medical Journal*, 325, 1312–13

Buss, D. M. (1989). Sex differences in human mate preferences: Evolutionary hypotheses tested in 37 cultures. *Behavioral and Brain Sciences*, 12, 1–49

Byers, J. et al. (2010). Female mate choice based upon male motor performance. *Animal Behavior*, 79, 771–8

Cabanac, M. (1979). Sensory pleasure. *Quarterly Review of Biology*, 54, 1–22

Cackowski, J. M. and Nasar, J. L. (2003). The restorative effects of roadside vegetation: Implications for automobile driver anger and frustration. *Environment and Behavior*, 35, 736–51

Cain, S. (2012). *Quiet: The power of introverts in a world that can't stop talking*. New York: Penguin

Calvert, G. A. et al. (eds.) (2004). *The handbook of multisensory processing*. Cambridge, MA: MIT Press

Camponogara, I. et al. (2017). Expert players accurately detect an opponent's movement intentions through sound alone. *Journal of Experimental Psychology: Human Perception and Performance*, 43, 348–59

Campos, C. et al. (2019). Dietary approaches to stop hypertension diet concordance and incident heart failure: The multi-ethnic study of atherosclerosis. *American Journal of Preventive Medicine*, 56, 89–96

Cañal-Bruland, R. et al. (2018). Auditory contributions to visual anticipation in tennis. *Psychology of Sport and Exercise*, 36, 100–103

Carlin, S. et al. (1962). Sound stimulation and its effect on dental sensation threshold. *Science*, 138, 1258–9

Carrus, G. et al. (2017). A different way to stay in touch with 'urban nature': The perceived restorative qualities of botanical gardens. *Frontiers in Psychology*, 8, 914

Carskadon, M. A. and Herz, R. S. (2004). Minimal olfactory perception during sleep: Why odor alarms will not work for humans. *Sleep*, 27, 402–5

Carter, J. M. et al. (2004). The effect of glucose infusion on glucose kinetics during a 1-h time trial. *Medicine and Science in Sports and Exercise*, 36, 1543–50

Castiello, U. et al. (2006). Cross-modal interactions between olfaction and vision when grasping. *Chemical Senses*, 31, 665–71

Chambers, E. S. et al. (2009). Carbohydrate sensing in the human mouth: Effects on exercise performance and brain activity. *Journal of Physiology*, 587, 1779–94

Chang, A.-M. et al. (2015). Evening use of light-emitting eReaders negatively affects sleep, circadian timing, and next-morning alertness. *Proceedings of the National Academy of Sciences of the USA*, 112, 1232–7

Chang, T. Y. and Kajackaite, A. (2019). Battle for the thermostat: Gender and the effect of temperature on cognitive performance. *PLOS One*, 14, e0216362

Changizi, M. A. et al. (2006). Bare skin, blood and the evolution of primate colour vision. *Biology Letters*, 2, 217–21

Charlton, B. D. et al. (2012). Do women prefer more complex music around ovulation? *PLOS One*, 7, e35626

Chekroud, S. R. et al. (2018). Association between physical exercise and mental health in 1.2 million individuals in the USA between 2011 and 2015: A cross-sectional study. *Lancet Psychiatry*, 5, 739–46

Chellappa, S. L. et al. (2011). Can light make us bright? Effects of light on cognition and sleep. *Progress in Brain Research*, 190, 119–33

Chen, X. et al. (2014). The moderating effect of stimulus attractiveness on the effect of alcohol consumption on attractiveness ratings. *Alcohol and Alcoholism*, 49, 515–19

Cheskin, L. and Ward, L. B. (1948). Indirect approach to market reactions. *Harvard Business Review*, 26, 572–80

Cho, S. et al. (2015). Blue lighting decreases the amount of food consumed in men, but not in women. *Appetite*, 85, 111–17

Churchill, A. et al. (2009). The cross-modal effect of fragrance in shampoo: Modifying the perceived feel of both product and hair during and after washing. *Food Quality and Preference*, 20, 320–28

Classen, C. (2012). *The deepest sense: A cultural history of touch*. Chicago: University of Illinois Press

Classen, C. et al. (1994). *Aroma: The cultural history of smell*. London: Routledge

Clifford, C. (1985). New scent waves. *Self*, December, 115–17

Cohen, B. et al. (1989). At the movies: An unobtrusive study of arousal-attraction. *Journal of Social Psychology*, 129, 691–3

Cohen, S. et al. (2015). Does hugging provide stress-buffering social support? A study of susceptibility to upper respiratory infection and illness. *Psychological Science*, 26, 135–47

Collins, J. F. (1965). The colour temperature of daylight. *British Journal of Applied Psychology*, 16, 527–32

Colvile, R. (2017). *The great acceleration: How the world is getting faster, faster*. London: Bloomsbury

Conrad, C. et al. (2007). Overture for growth hormone: Requiem for interleukin-6? *Critical Care Medicine*, 35, 2709–13

Corbin, A. (1986). *The foul and the fragrant: Odor and the French social imagination*. Cambridge, MA: Harvard University Press

Costa, M. et al. (2018). Interior color and psychological functioning in a university residence hall. *Frontiers in Psychology*, 9, 1580

Craig, R. et al. (2009). *Health survey for England 2008*, vol. 1: *Physical activity and fitness*. NHS Information Centre for Health and Social Care: Leeds, www.healthypeople.gov/2020/topics-objectives/topic/physical-activity

Crawford, I. (1997). *Sensual home: Liberate your senses and change your life*. London: Quadrille Publishing

Croon, E. et al. (2005). The effect of office concepts on worker health and performance: A systematic review of the literature. *Ergonomics*, 48, 119–34

Crossman, M. K. (2017). Effects of interactions with animals on human psychological distress. *Journal of Clinical Psychology*, 73, 761–84

Crowley, K. (2011). Sleep and sleep disorders in older adults. *Neuropsychology Review*, 21, 41–53

Croy, I. et al. (2015). Reduced pleasant touch appraisal in the presence of a disgusting odor. *PLOS One*, 9, e92975

Cummings, B. E. and Waring, M. S. (2020). Potted plants do not improve indoor air quality: a review and analysis of reported VOC removal efficiencies. *Journal of Exposure Science and Environmental Epidemiology*, 30, 253–61

Cutting, J. E. (2006). The mere exposure effect and aesthetic preference. In P. Locher et al. (eds.), *New directions in aesthetics, creativity, and the arts*. Amityville, NY: Baywood Publishing, pp. 33–46

Czeisler, C. A. et al. (1986). Bright light resets the human circadian pacemaker independent of the timing of the sleep-wake cycle. *Science*, 233, 667–71

Dalke, H. et al. (2006). Colour and lighting in hospital design. *Optics and Laser Technology*, 38, 343–65

Dalton, P. (1996). Odor perception and beliefs about risk. *Chemical Senses*, 21, 447–58

Dalton, P. and Wysocki, C. J. (1996). The nature and duration of adaptation following long-term odor exposure. *Perception and Psychophysics*, 58, 781–92

Darbyshire, J. L. (2016). Excessive noise in intensive care units. *British Medical Journal*, 353, i1956

Darbyshire, J. L. and Young, J. D. (2013). An investigation of sound levels on intensive care units with reference to the WHO guidelines. *Critical Care*, 17, R187

Darwin, C. (1871). The descent of man, and selection in relation to sex. In E. O. Wilson (ed.) (2006), *From so simple a beginning: The four great books of Charles Darwin*. New York: W. W. Norton

Dazkir, S. S. and Read, M. A. (2012). Furniture forms and their influence on our emotional responses toward interior environments. *Environment and Behavior*, 44, 722–34

de Bell, S. et al. (2020). Spending time in the garden is positively associated with health and wellbeing: Results from a national survey in England. *Landscape and Urban Planning*, 200, 103836.

de Wijk, R. A. et al. (2018). Supermarket shopper movements versus sales, and the effects of scent, light, and sound. *Food Quality and Preference*, 68, 304–14

Dematté, M. L. et al. (2006). Cross-modal interactions between olfaction and touch. *Chemical Senses*, 31, 291–300

Dematté, M. L. et al. (2007). Olfactory cues modulate judgments of facial attractiveness. *Chemical Senses*, 32, 603–10

Denworth, L. (2015). The social power of touch. *Scientific American Mind*, July/August, 30–39

Diaconu, M. et al. (eds.) (2011). *Senses and the city: An interdisciplinary approach to urban sensescapes*. Vienna, Austria: Lit Verlag

Diamond, J. (1993). New Guineans and their natural world. In S. R. Kellert and E. O. Wilson (eds.), *The biophilia hypothesis*. Washington, DC: Island Press, pp. 251–74

Diette, G. B. et al. (2003). Distraction therapy with nature sights and sounds reduces pain during flexible bronchoscopy: A complementary approach to routine analgesia. *Chest*, 123, 941–8

Dijkstra, K. et al. (2008). Stress-reducing effects of indoor plants in the built healthcare environment: The mediating role of perceived attractiveness. *Preventative Medicine*, 47, 279–83

Dobzhansky, T. (1973). Nothing in biology makes sense except in the light of evolution. *American Biology Teacher*, 35, 125–9

Dolan, B. (2004). *Josiah Wedgwood: Entrepreneur to the enlightenment*. London: HarperPerennial

Dunn, W. (2007). *Living sensationally: Understanding your senses*. London: Jessica Kingsley

Dutton, D. G. and Aron, A. P. (1974). Some evidence for heightened sexual attraction under conditions of high anxiety. *Journal of Personality and Social Psychology*, 30, 510–17

Edworthy, J. and Waring, H. (2006). The effects of music tempo and loudness level on treadmill exercise. *Ergonomics*, 49, 1597–610

Einöther, S. J. and Martens, V. E. (2013). Acute effects of tea consumption on attention and mood. *American Journal of Clinical Nutrition*, 98, 1700S–1708S

Ellingsen, D.-M. et al. (2016). The neurobiology shaping affective touch: Expectation, motivation, and meaning in the multisensory context. *Frontiers in Psychology*, 6, 1986

Elliot, A. J. and Niesta, D. (2008). Romantic red: Red enhances men's attraction to women. *Journal of Personality and Social Psychology*, 95, 1150–64

Elliot, A. J. and Pazda, A. D. (2012). Dressed for sex: Red as a female sexual signal in humans. *PLOS One*, 7, e34607

Elliot, A. J. et al. (2007). Color and psychological functioning: The effect of red on performance attainment. *Journal of Experimental Psychology: General*, 136, 154–68

Etzi, R. et al. (2014). Textures that we like to touch: An experimental study of aesthetic preferences for tactile stimuli. *Consciousness and Cognition*, 29, 178–88

Evans, D. (2002). *Emotion: The science of sentiment*. Oxford: Oxford University Press

Evans, G. W. and Johnson, D. (2000). Stress and open-office noise. *Journal of Applied Psychology*, 85, 779–83

Evans, W. N. and Graham, J. D. (1991). Risk reduction or risk compensation? The case of mandatory safety-belt use laws. *Journal of Risk and Uncertainty*, 4, 61–73

Facer-Childs, E. R. et al. (2019). Resetting the late timing of 'night owls' has a positive impact on mental health and performance. *Sleep Medicine*, 60, 236–47

Fancourt, D. et al. (2016). The razor's edge: Australian rock music impairs men's performance when pretending to be a surgeon. *Medical Journal of Australia*, 205, 515–18

Feinberg, D. R. et al. (2008). Correlated preferences for men's facial and vocal masculinity. *Evolution and Human Behavior*, 29, 233–41

Feldstein, I. T. and Peli, E. (2020). Pedestrians accept shorter distances to light vehicles than dark ones when crossing the street. *Perception*, 49, 558–66

Fenko, A. and Loock, C. (2014). The influence of ambient scent and music on patients' anxiety in a waiting room of a plastic surgeon. *HERD: Health Environments Research and Design Journal*, 7, 38–59

Fich, L. B. et al. (2014). Can architectural design alter the physiological reaction to psychosocial stress? A virtual TSST experiment. *Physiology and Behavior*, 135, 91–7

Field, T. (2001). *Touch*. Cambridge, MA: MIT Press

Field, T. et al. (1996). Massage therapy reduces anxiety and enhances EEG pattern of alertness and math computations. *International Journal of Neuroscience*, 86, 197–205

Field, T. et al. (2008). Lavender bath oil reduces stress and crying and enhances sleep in very young infants. *Early Human Development*, 84, 399–401

Fisk, W. J. (2000). Health and productivity gains from better indoor environments and their relationship with building energy efficiency. *Annual Review of Energy and the Environment*, 25, 537–66

Fismer, K. L. and Pilkington, K. (2012). Lavender and sleep: A systematic review of the evidence. *European Journal of Integrative Medicine*, 4, e436–e447

Forster, S. and Spence, C. (2018). 'What smell?' Temporarily loading visual attention induces prolonged inattentional anosmia. *Psychological Science*, 29, 1642–52

Fox, J. G. and Embrey, E. D. (1972). Music: An aid to productivity. *Applied Ergonomics*, 3, 202–5

Frank, M. G. and Gilovich, T. (1988). The dark side of self- and social perception: Black uniforms and aggression in professional sports. *Journal of Personality and Social Psychology*, 54, 74–85

Franklin, D. (2012). How hospital gardens help patients heal. *Scientific American*, 1 March, www.scientificamerican.com/article/nature-that-nurtures/

Fritz, T. H. et al. (2013). Musical agency reduces perceived exertion during strenuous physical performance. *Proceedings of the National Academy of Sciences of the USA*, 110, 17784–9

Fruhata, T. et al. (2013). Doze sleepy driving prevention system (finger massage, high density oxygen spray, grapefruit fragrance) with that involves chewing dried shredded squid. *Procedia Computer Science*, 22, 790–99

Frumkin, H. (2001). Beyond toxicity: Human health and the natural environment. *American Journal of Preventative Medicine*, 20, 234–40

Fukuda, M. and Aoyama, K. (2017). Decaffeinated coffee induces a faster conditioned reaction time even when participants know that the drink does not contain caffeine. *Learning and Motivation*, 59, 11–18

Fuller, R. A. and Gaston, K. J. (2009). The scaling of green space coverage in European cities. *Biology Letters*, 5, 352–5

Fuller, R. A. et al. (2007). Psychological benefits of greenspace increase with biodiversity. *Biology Letters*, 3, 390–94

Fumento, M. (1998). 'Road rage' versus reality. *Atlantic Monthly*, 282, 12–17

Gabel, V. et al. (2013). Effects of artificial dawn and morning blue light on daytime cognitive performance, well-being, cortisol and melatonin levels. *Chronobiology International*, 30, 988–97

Gafsou, M. and Hildyard, D. (2019). H+. *Granta*, 148, 94–128

Gallace, A. and Spence, C. (2014). *In touch with the future: The sense of touch from cognitive neuroscience to virtual reality*. Oxford: Oxford University Press

Galton, F. (1883). *Inquiries into human faculty and its development*. London: Macmillan

García-Segovia, P. et al. (2015). Influence of table setting and eating location on food acceptance and intake. *Food Quality and Preference*, 39, 1–7

Gatti, M. F. and da Silva, M. J. P. (2007). Ambient music in emergency services: The professionals' perspective. *Latin American Journal of Nursing*, 15, 377–83

Geddes, L. (2020). How to hug people in a coronavirus-stricken world. *New Scientist*, 5 August, www.newscientist.com/article/mg24732944-300-how-to-hug-people-in-a-coronavirus-stricken-world/#ixzz6UKxBNFzI

Genschow, O. et al. (2015). Does Baker-Miller pink reduce aggression in prison detention cells? A critical empirical examination. *Psychology, Crime and Law*, 21, 482–9

Geschwind, N. and Galaburda, A. M. (1985). Cerebral lateralization. Biological mechanisms, associations, and pathology: A hypothesis and a program for research. *Archives of Neurology*, 42, 428–59, 521–52, 634–54

Gibson, J. J. and Crooks, L. E. (1938). A theoretical field-analysis of automobile-driving. *American Journal of Psychology*, 51, 453–71

Gibson, M. and Shrader, J. (2014). Time use and productivity: The wage returns to sleep. UC San Diego Department of Economics Working Paper

Gillis, K. and Gatersleben, B. (2015). A review of psychological literature on the health and wellbeing benefits of biophilic design. *Buildings*, 5, 948–63

Glacken, C. J. (1967). *Traces on the Rhodian shore: Nature and culture in Western thought from ancient times to the end of the Eighteenth Century*. Berkeley, CA: University of California Press

Gladue, B. and Delaney, H. J. (1990). Gender differences in perception of attractiveness of men and women in bars. *Personality and Social Psychology Bulletin*, 16, 378–91

Glass, S. T. et al. (2014). Do ambient urban odors evoke basic emotions? *Frontiers in Psychology*, 5, 340

Golan, A. and Fenko, A. (2015). Toward a sustainable faucet design: Effects of sound and vision on perception of running water. *Environment and Behavior*, 47, 85–101

Goldstein, P. et al. (2017). The role of touch in regulating inter-partner physiological coupling during empathy for pain. *Scientific Reports*, 7, 3252

Gori, M. et al. (2008). Young children do not integrate visual and haptic information. *Current Biology*, 18, 694–8

Graff, V. et al. (2019). Music versus midazolam during preoperative nerve block placements: A prospective randomized controlled study. *Regional Anesthesia and Pain Medicine*, 44, 796–9

Graham-Rowe, D. (2001). Asleep at the wheel. *New Scientist*, 169, 24

Grammer, K. et al. (2004). Disco clothing, female sexual motivation, and relationship status: Is she dressed to impress? *Journal of Sex Research*, 41, 66–74

Greene, M. R. and Oliva, A. (2009). The briefest of glances: The time course of natural scene understanding. *Psychological Science*, 20, 464–72

Greenfield, A. B. (2005). *A perfect red: Empire, espionage, and the quest for the color of desire.* New York: HarperCollins

Griskevicius, V. and Kenrick, D. T. (2013). Fundamental motives: How evolutionary needs influence consumer behavior. *Journal of Consumer Psychology*, 23, 372–86

Groyecka, A. et al. (2017). Attractiveness is multimodal: Beauty is also in the nose and ear of the beholder. *Frontiers in Psychology*, 8, 778

Gubbels, J. L. (1938). *American highways and roadsides.* Boston, MA: Houghton-Mifflin

Guéguen, N. (2012). Color and women attractiveness: When red clothed women are perceived to have more intense sexual intent. *Journal of Social Psychology*, 152, 261–5

Guéguen, N. and Jacob, C. (2011). Enhanced female attractiveness with use of cosmetics and male tipping behavior in restaurants. *Journal of Cosmetic Science*, 62, 283–90

——— (2014). Clothing color and tipping: Gentlemen patrons give more tips to waitresses with red clothes. *Journal of Hospitality and Tourism Research*, 38, 275–80

Guéguen, N. et al. (2012). When drivers see red: Car color frustrators and drivers' aggressiveness. *Aggressive Behaviour*, 38, 166–9

Guieysse, B. et al. (2008). Biological treatment of indoor air for VOC removal: Potential and challenges. *Biotechnology Advances*, 26, 398–410

Gupta, A. et al. (2018). Innovative technology using virtual reality in the treatment of pain: Does it reduce pain via distraction, or is there more to it? *Pain Medicine*, 19, 151–9

Haehner, A. et al. (2017). Influence of room fragrance on attention, anxiety and mood. *Flavour and Fragrance Journal*, 1, 24–8

Hafner, M. et al. (2016). Why sleep matters – the economic costs of insufficient sleep. A cross-country comparative analysis. Rand Corporation, www.rand.org/pubs/research_reports/RR1791.html

Haga, A. et al. (2016). Psychological restoration can depend on stimulus-source attribution: A challenge for the evolutionary account. *Frontiers in Psychology*, 7, 1831

Hagemann, N. et al. (2008). When the referee sees red. *Psychological Science*, 19, 769–71

Hagerhall, C. M. et al. (2004). Fractal dimension of landscape silhouette outlines as a predictor of landscape preference. *Journal of Environmental Psychology*, 24, 247–55

Haghayegh, S. et al. (2019). Before-bedtime passive body heating by warm shower or bath to improve sleep: A systematic review and meta-analysis. *Sleep Medicine Reviews*, 46, 124–35

Hamilton, A. (1966). What science is learning about smell. *Science Digest*, 55 (November), 81–4

Han, K. (2007). Responses to six major terrestrial biomes in terms of scenic beauty, preference, and restorativeness. *Environment and Behavior*, 39, 529–56

Hanss, D. et al. (2012). Active red sports car and relaxed purple-blue van: Affective qualities predict color appropriateness for car types. *Journal of Consumer Behaviour*, 11, 368–80

Harada, H. et al. (2018). Linalool odor-induced anxiolytic effects in mice. *Frontiers in Behavioral Neuroscience*, 12, 241

Hardy, M. et al. (1995). Replacement of drug treatment for insomnia by ambient odour. *The Lancet*, 346, 701

Harlow, H. F. and Zimmerman, R. R. (1959). Affectional responses in the infant monkey. *Science*, 130, 421–32

Harper, M. B. et al. (2015). Photographic art in exam rooms may reduce white coat hypertension. *Medical Humanities*, 41, 86–8

Hartig, T. et al. (2011). Health benefits of nature experience: Psychological, social and cultural processes. In K. Nilsson et al. (eds.), *Forests, trees and human health*. Berlin: Springer Science, pp. 127–68

Harvey, A. G. (2003). The attempted suppression of presleep cognitive activity in insomnia. *Cognitive Therapy and Research*, 27, 593–602

Harvey, A. G. and Payne, S. (2002). The management of unwanted pre-sleep thoughts in insomnia: Distraction with imagery versus general distraction. *Behaviour Research and Therapy*, 40, 267–77

Haslam, S. A. and Knight, C. (2010). Cubicle, sweet cubicle. *Scientific American Mind*, September/October, 30–35

Haverkamp, M. (2014). *Synesthetic design: Handbook for a multisensory approach*. Basel: Birkhäuser

Haviland-Jones, J. et al. (2005). An environmental approach to positive emotion: Flowers. *Evolutionary Psychology*, 3, 104–32

Havlíček, J. et al. (2006). Non-advertised does not mean concealed: Body odour changes across the human menstrual cycle. *Ethology*, 112, 81–90

Havlíček, J. et al. (2008). He sees, she smells? Male and female reports of sensory reliance in mate choice and non-mate choice contexts. *Personality and Individual Differences*, 45, 565–70

Hawksworth, C. et al. (1997). Music in theatre: Not so harmonious. A survey of attitudes to music played in the operating theatre. *Anaesthesia*, 52, 79–83

Hedblom, M. et al. (2014). Bird song diversity influences young people's appreciation of urban landscapes. *Urban Forestry and Urban Greening*, 13, 469–74

Hellier, E. et al. (2011). The influence of auditory feedback on speed choice, violations and comfort in a driving simulation game. *Transportation Research Part F: Traffic Psychology and Behaviour*, 14, 591–9

Helmefalk, M. and Hultén, B. (2017). Multi-sensory congruent cues in designing retail store atmosphere: Effects on shoppers' emotions and purchase behaviour. *Journal of Retailing and Consumer Services*, 38, 1–11

Hepper, P. G. (1988). Fetal 'soap' addiction. *The Lancet*, 11 June, 1347–8

Herz, R. (2007). *The scent of desire: Discovering our enigmatic sense of smell*. New York: William Morrow

——— (2009). Aromatherapy facts and fictions: A scientific analysis of olfactory effects on mood, physiology and psychology. *International Journal of Neuroscience*, 119, 263–90

Herz, R. S. and Cahill, E. D. (1997). Differential use of sensory information in sexual behavior as a function of gender. *Human Nature*, 8, 275–86

Heschong, L. (1979). *Thermal delight in architecture*. Cambridge, MA: MIT Press

Hewlett, S. A. and Luce, C. B. (2006). Extreme jobs: The dangerous allure of the 70-hour workweek. *Harvard Business Review*, December, https://hbr.org/2006/12/extreme-jobs-the-dangerous-allure-of-the-70-hour-workweek

Higham, W. (2019). *The work colleague of the future: A report on the long-term health of office workers*. Report commissioned by Fellowes, July, https://assets.fellowes.com/skins/fellowes/responsive/gb/en/resources/work-colleague-of-the-future/download/WCOF_Report_EU.pdf

Hilditch, C. J. et al. (2016). Time to wake up: Reactive countermeasures to sleep inertia. *Industrial Health*, 54, 528–41

Hill, A. W. (1915). The history and functions of botanic gardens. *Annals of the Missouri Botanical Garden*, 2, 185–240

Hill, R. A. and Barton, R. A. (2005). Red enhances human performance in contests. *Nature*, 435, 293

Hillman, C. H. et al. (2008). Be smart, exercise your heart: Exercise effects on brain and cognition. *Nature Reviews Neuroscience*, 9, 58–65

Hilton, K. (2015). Psychology: The science of sensory marketing. *Harvard Business Review*, March, 28–31, https://hbr.org/2015/03/the-science-of-sensory-marketing

Hirano, H. (1996). *5 pillars of the visual workplace: The sourcebook for 5S implementation*. New York: Productivity Press

Ho, C. and Spence, C. (2005). Olfactory facilitation of dual-task performance. *Neuroscience Letters*, 389, 35–40

——— (2008). *The multisensory driver: Implications for ergonomic car interface design*. Aldershot: Ashgate

——— (2009). Using peripersonal warning signals to orient a driver's gaze. *Human Factors*, 51, 539–56

——— (2013). Affective multisensory driver interface design. *International Journal of Vehicle Noise and Vibration* (Special Issue on *Human Emotional Responses to Sound and Vibration in Automobiles*), 9, 61–74

Hoehl, S. et al. (2017). Itsy bitsy spider: Infants react with increased arousal to spiders and snakes. *Frontiers in Psychology*, 8, 1710

Hoekstra, S. P. et al. (2018). Acute and chronic effects of hot water immersion on inflammation and metabolism in sedentary, overweight adults. *Journal of Applied Physiology*, 125, 2008–18

Holgate, S. T. (2017). 'Every breath we take: The lifelong impact of air pollution' – a call for action. *Clinical Medicine*, 17, 8–12

Holland, R. W. et al. (2005). Smells like clean spirit. Nonconscious effects of scent on cognition and behavior. *Psychological Science*, 16, 689–93

Hollingworth, H. L. (1939). Chewing as a technique of relaxation. *Science*, 90, 385–7

Holmes, C. et al. (2002). Lavender oil as a treatment for agitated behaviour in severe dementia: A placebo controlled study. *International Journal of Geriatric Psychiatry*, 17, 305–8

Homburg, C. et al. (2012). Of dollars and scents – Does multisensory marketing pay off? Working paper, Institute for Marketing Oriented Management.

Hongisto, V. et al. (2017). Perception of water-based masking sounds – long-term experiment in an open-plan office. *Frontiers in Psychology*, 8, 1177

Horswill, M. S. and Plooy, A. M. (2008). Auditory feedback influences perceived driving speeds. *Perception*, 37, 1037–43

Hove, M. J. and Risen, J. L. (2009). It's all in the timing: Interpersonal synchrony increases affiliation. *Social Cognition*, 27, 949–61

Howes, D. (ed.) (2004). *Empire of the senses: The sensual culture reader.* Oxford: Berg

——— (2014). *A cultural history of the senses in the modern age.* London: Bloomsbury Academic

Howes, D. and Classen, C. (2014). *Ways of sensing: Understanding the senses in society.* London: Routledge

Huang, L. et al. (2011). Powerful postures versus powerful roles: Which is the proximate correlate of thought and behaviour? *Psychological Science*, 22, 95–102

Hugill, N. et al. (2010). The role of human body movements in mate selection. *Evolutionary Psychology*, 8, 66–89

Hull, J. M. (1990). *Touching the rock: An experience of blindness.* London: Society for Promoting Christian Knowledge

Hulsegge, J. and Verheul, A. (1987). *Snoezelen: another world. A practical book of sensory experience environments for the mentally handicapped.* Chesterfield: ROMPA

Hultén, B. (2012). Sensory cues and shoppers' touching behaviour: The case of IKEA. *International Journal of Retail and Distribution Management*, 40, 273–89

Huss, E. et al. (2018). Humans' relationship to flowers as an example of the multiple components of embodied aesthetics. *Behavioral Sciences*, 8, 32

Hutmacher, F. (2019). Why is there so much more research on vision than on any other sensory modality? *Frontiers in Psychology*, 10, 2246

Huxley, A. (1954). *The doors of perception.* London: Harper & Brothers

Imschloss, M. and Kuehnl, C. (2019). Feel the music! Exploring the cross-modal correspondence between music and haptic perceptions of softness. *Journal of Retailing*, 95, 158–69

Itten, J. and Birren, F. (1970). *The elements of color* (trans. E. van Hagen). New York: John Wiley & Sons

Jacob, C. et al. (2012). She wore something in her hair: The effect of ornamentation on tipping. *Journal of Hospitality Marketing and Management*, 21, 414–20

Jacobs, K. W. and Hustmyer, F. E. (1974). Effects of four psychological primary colors on GSR, heart rate and respiration rate. *Perceptual and Motor Skills*, 38, 763–6

Jacquier, C. and Giboreau, A. (2012). Perception and emotions of colored atmospheres at the restaurant. *Predicting Perceptions: Proceedings of the 3rd International Conference on Appearance*, pp. 165–7

James, L. and Nahl, D. (2000). *Road rage*. Amherst, NY: Prometheus Books

James, W. (1890). *The principles of psychology* (2 vols.). New York: Henry Holt

Jewett, M. E. et al. (1999). Time course of sleep inertia dissipation in human performance and alertness. *Journal of Sleep Research*, 8, 1–8

Jones, A. L. and Kramer, R. S. S. (2016). Facial cosmetics and attractiveness: Comparing the effect sizes of professionally-applied cosmetics and identity. *PLOS One*, 11, e0164218

Jones, A. L. et al. (2018). Positive facial affect looks healthy. *Visual Cognition*, 26, 1–12

Jones, B. T. et al. (2003). Alcohol consumption increases attractiveness ratings of opposite-sex faces: A possible third route to risky sex. *Addiction*, 98, 1069–75

Jones, S. E. et al. (2019). Genome-wide association analyses of chronotype in 697,828 individuals provides insights into circadian rhythms. *Nature Communications*, 10, 343

Joye, Y. (2007). Architectural lessons from environmental psychology: The case of biophilic architecture. *Review of General Psychology*, 11, 305–28

Joye, Y. and van den Berg, A. (2011). Is love for green in our genes? A critical analysis of evolutionary assumptions in restorative environments research. *Urban Forestry and Urban Greening*, 10, 261–8

Just, M. G. et al. (2019). Human indoor climate preferences approximate specific geographies. *Royal Society Open Science*, 6, 180695

Jütte, R. (2005). *A history of the senses: From antiquity to cyberspace.* Cambridge: Polity Press

Kabat-Zinn, J. (2005). *Coming to our senses: Healing ourselves and the world through mindfulness.* New York: Hyperion

Kahn, P. H., Jr (1999). *The human relationship with nature: Development and culture.* Cambridge, MA: MIT Press

Kahn, P. H., Jr et al. (2008). A plasma display window? The shifting baseline problem in a technologically mediated natural world. *Journal of Environmental Psychology*, 28, 192–9

Kahn, P. H., Jr et al. (2009). The human relation with nature and technological nature. *Current Directions in Psychological Science*, 18, 37–42

Kaida, K., et al. (2006). Indoor exposure to natural bright light prevents afternoon sleepiness. *Sleep*, 29, 462–9

Kampe, K. K. et al. (2001). Reward value of attractiveness and gaze. *Nature*, 413, 589

Kampfer, K. et al. (2017). Touch-flavor transference: Assessing the effect of packaging weight on gustatory evaluations, desire for food and beverages, and willingness to pay. *PLOS One*, 12(10), e0186121

Kaplan, K. A. et al. (2019). Effect of light flashes vs sham therapy during sleep with adjunct cognitive behavioral therapy on sleep quality among adolescents: A randomized clinical trial. *JAMA Network Open*, 2, e1911944

Kaplan, R. (1973). Some psychological benefits of gardening. *Environment and Behavior*, 5, 145–52

Kaplan, R. and Kaplan, S. (1989). *The experience of nature: A psychological perspective.* New York: Cambridge University Press

Kaplan, S. (1995). The restorative benefits of nature: Toward an integrative framework. *Journal of Environmental Psychology*, 15, 169–82

—— (2001). Meditation, restoration, and the management of mental fatigue. *Environment and Behavior*, 33, 480–506

Karageorghis, C. I. and Terry, P. C. (1997). The psychophysical effects of music in sport and exercise: A review. *Journal of Sport Behavior*, 20, 54–168

Karim, A. A. et al. (2017). Why is 10 past 10 the default setting for clocks and watches in advertisements? A psychological experiment. *Frontiers in Psychology*, 8, 1410

Karremans, J. C. et al. (2006). Beyond Vicary's fantasies: The impact of subliminal priming and branded choice. *Journal of Experimental Social Psychology*, 42, 792–8

Katz, J. (2014). Noise in the operating room. *Anesthesiology*, 121, 894–9

Keller, A. (2008). Toward the dominance of vision? *Science*, 320, 319

Kellert, S. R. and Wilson, E. O. (eds.) (1993). *The biophilia hypothesis*. Washington, DC: Island Press

Kingma, B. and van Marken Lichtenbelt, W. D. (2015). Energy consumption in buildings and female thermal demand. *Nature Climate Change*, 5, 1054–6

Kirk-Smith, M. (2003). The psychological effects of lavender 1: In literature and plays. *International Journal of Aromatherapy*, 13, 18–22

Knasko, S. C. (1989). Ambient odor and shopping behavior. *Chemical Senses*, 14, 718

Kniffin, K. M. et al. (2015). Eating together at the firehouse: How workplace commensality relates to the performance of firefighters. *Human Performance*, 28, 281–306

Knight, C. and Haslam, S. A. (2010). The relative merits of lean, enriched, and empowered offices: An experimental examination of the impact of workspace management. *Journal of Experimental Psychology: Applied*, 16, 158–72

Knoeferle, K. et al. (2012). It is all in the mix: The interactive effect of music tempo and mode on in-store sales. *Marketing Letters*, 23, 325–37

Knoeferle, K. et al. (2016). Multisensory brand search: How the meaning of sounds guides consumers' visual attention. *Journal of Experimental Psychology: Applied*, 22, 196–210

Knopf, R. C. (1987). Human behavior, cognition, and affect in the natural environment. In D. Stokols and I. Altman (eds.), *Handbook of environmental psychology*, vol. 1. New York: John Wiley & Sons, pp. 783–825

Kochanek, K. D. et al. (2014). Mortality in the United States, 2013. *NCHS Data Brief*, 178, 1–8

Koga, K. and Iwasaki, Y. (2013). Psychological and physiological effect in humans of touching plant foliage – using the semantic differential method and cerebral activity as indicators. *Journal of Physiological Anthropology*, 32, 7

Kohara, K. et al. (2018). Habitual hot water bathing protects cardiovascular function in middle-aged to elderly Japanese subjects. *Scientific Reports*, 8, 8687

Körber, M. et al. (2015). Vigilance decrement and passive fatigue caused by monotony in automated driving. *Procedia Manufacturing*, 3, 2403–9

Kort, Y. A. W. et al. (2006). What's wrong with virtual trees? Restoring from stress in a mediated environment. *Journal of Environmental Psychology*, 26, 309–20

Kotler, P. (1974). Atmospherics as a marketing tool. *Journal of Retailing*, 49, 48–64

Kozusznik, M. W. et al. (2019). Decoupling office energy efficiency from employees' well-being and performance: A systematic review. *Frontiers in Psychology*, 10, 293

Kranowitz, C. S. (1998). *The out-of-sync child: Recognizing and coping with sensory integration*. New York: Penguin Putnam

Kräuchi, K. et al. (1999). Warm feet promote the rapid onset of sleep. *Nature*, 401, 36–7

Kreutz, G. et al. (2018). In dubio pro silentio – Even loud music does not facilitate strenuous ergometer exercise. *Frontiers in Psychology*, 9, 590

Krieger, M. H. (1973). What's wrong with plastic trees? Artifice and authenticity in design. *Science*, 179, 446–55

Kripke, D. F. et al. (2012). Hypnotics' association with mortality or cancer: A matched cohort study. *BMJ Open*, 2, e000850

Kühn, S. et al. (2016). Multiple 'buy buttons' in the brain: Forecasting chocolate sales at point-of-sale based on functional brain activation using fMRI. *NeuroImage*, 136, 122–8

Kühn, S. et al. (2017). In search of features that constitute an 'enriched environment' in humans: Associations between geographical properties and brain structure. *Scientific Reports*, 7, 11920

Küller, R. et al. (2006). The impact of light and colour on psychological mood: A cross-cultural study of indoor work environments. *Ergonomics*, 49, 1496–507

Kunst-Wilson, W. R. and Zajonc, R. B. (1980). Affective discrimination of stimuli that cannot be recognized. *Science*, 207, 557–8

Kurzweil, R. (2005). *The singularity is near: When humans transcend biology*. London: Prelude

Kuukasjärvi, S. et al. (2004). Attractiveness of women's body odors over the menstrual cycle: The role of oral contraceptives and receiver sex. *Behavioral Ecology*, 15, 579–84

Kwallek, N. and Lewis, C. M. (1990). Effects of environmental colour on males and females: A red or white or green office. *Applied Ergonomics*, 21, 275–8

Kwallek, N. et al. (1996). Effects of nine monochromatic office interior colors on clerical tasks and worker mood. *Color Research and Application*, 21, 448–58

Kweon, B.-S. et al. (2008). Anger and stress: The role of landscape posters in an office setting. *Environment and Behavior*, 40, 355–81

Kyle, S. D. et al. (2010). '. . . Not just a minor thing, it is something major, which stops you from functioning daily': Quality of life and daytime functioning in insomnia. *Behavioral Sleep Medicine*, 8, 123–40

Lamote de Grignon Pérez, J. et al. (2019). Sleep differences in the UK between 1974 and 2015: Insights from detailed time diaries. *Journal of Sleep Research*, 28, e12753

Lankston, L. et al. (2010). Visual art in hospitals: Case studies and review of the evidence. *Journal of the Royal Society of Medicine*, 103, 490–99

Lanza, J. (2004). *Elevator music: A surreal history of Muzak, easy-listening, and other moodsong.* Ann Arbor: University of Michigan Press

Lay, M. G. (1992). *Ways of the world: A history of the world's roads and of the vehicles that used them.* New Brunswick, NJ: Rutgers University Press

Le Breton, D. (2017). *Sensing the world: An anthropology of the senses* (trans. C. Ruschiensky). London: Bloomsbury

Le Corbusier (1948/1972). *Towards a new architecture* (trans. F. Etchells). London: The Architectural Press

—— (1987). *The decorative art of today* (trans. J. L. Dunnett). Cambridge, MA: MIT Press

Leather, P. et al. (1998). Windows in the workplace: Sunlight, view, and occupational stress. *Environment and Behavior*, 30, 739–62

Lee, I. F. (2018). *Joyful: The surprising power of ordinary things to create extraordinary happiness.* London: Rider

Lee, K. E. et al. (2015). 40-second green roof views sustain attention: The role of micro-breaks in attention restoration. *Journal of Environmental Psychology*, 42, 182–9

Lee, R. and DeVore, I. (1968). *Man the hunter.* Chicago: Aldine

Leenders, M. A. A. M. et al. (2019). Ambient scent as a mood inducer in supermarkets: The role of scent intensity and time-pressure of shoppers. *Journal of Retailing and Consumer Services*, 48, 270–80

Lehrl, S. et al. (2007). Blue light improves cognitive performance. *Journal of Neural Transmission*, 114, 1435–63

Lehrner, J. et al. (2000). Ambient odor of orange in a dental office reduces anxiety and improves mood in female patients. *Physiology and Behavior*, 71, 83–6

Lenochová, P. et al. (2012). Psychology of fragrance use: Perception of individual odor and perfume blends reveals a mechanism for idiosyncratic effects on fragrance choice. *PLOS One*, 7, e33810

Levin, M. D. (1993). *Modernity and the hegemony of vision*. Berkeley: University of California Press

Levitin, D. (2015). *The organized mind: thinking straight in the age of information overload*. London: Penguin.

Lewis, D. M. G. et al. (2015). Lumbar curvature: A previously undiscovered standard of attractiveness. *Evolution and Human Behavior*, 36, 345–50

Lewis, D. M. G. et al. (2017). Why women wear high heels: Evolution, lumbar curvature, and attractiveness. *Frontiers in Psychology*, 8, 1875

Li, A. et al. (2011). Virtual reality and pain management: Current trends and future directions. *Pain Management*, 1, 147–57

Li, Q. (2010). Effect of forest bathing trips on human immune function. *Environmental Health and Preventative Medicine*, 15, 1, 9–17

Li, W. et al. (2007). Subliminal smells can guide social preferences. *Psychological Science*, 18, 1044–9

Lies, S. and Zhang, A. (2015). Prospective randomized study of the effect of music on the efficiency of surgical closures. *Aesthetic Surgery Journal*, 35, 858–63

Lin, H. (2014). Red-colored products enhance the attractiveness of women. *Displays*, 35, 202–5

Lindstrom, M. (2005). *Brand sense: How to build brands through touch, taste, smell, sight and sound*. London: Kogan Page

Liu, B. et al. (2015). Does happiness itself directly affect mortality? The prospective UK Million Women Study. *The Lancet*, 387, 874–81

Liu, J. et al. (2019). The impact of tablecloth on consumers' food perception in real-life eating situation. *Food Quality and Preference*, 71, 168–71

Lobmaier, J. S. et al. (2018). The scent of attractiveness: Levels of reproductive hormones explain individual differences in women's body odour. *Proceedings of the Royal Society B: Biological Sciences*, 285, 20181520

LoBue, V. (2014). Deconstructing the snake: The relative roles of perception, cognition, and emotion on threat detection. *Emotion*, 14, 701–11

Lockley, S. W. et al. (2006). Short-wavelength sensitivity for the direct effects of light on alertness, vigilance, and the waking electroencephalogram in humans. *Sleep*, 29, 161–8

Louv, R. (2005). *Last child in the woods: Saving our children from nature-deficit disorder*. Chapel Hill, NC: Algonquin Books

Lovato, N. and Lack, L. (2016). Circadian phase delay using the newly developed re-timer portable light device. *Sleep and Biological Rhythms*, 14, 157–64

Lupton, E. and Lipps, A. (eds.) (2018). *The senses: Design beyond vision*. Hudson, NY: Princeton Architectural Press

Lynn, M. et al. (2016). Clothing color and tipping: An attempted replication and extension. *Journal of Hospitality and Tourism Research*, 40, 516–24

Mace, B. L. et al. (1999). Aesthetic, affective, and cognitive effects of noise on natural landscape assessment. *Society and Natural Resources*, 12, 225–42

Mackerron, G. and Mourato, S. (2013). Happiness is greater in natural environments. *Global Environmental Change*, 23, 992–1000

Madzharov, A. et al. (2018). The impact of coffee-like scent on expectations and performance. *Journal of Environmental Psychology*, 57, 83–6

Malhotra, N. K. (1984). Information and sensory overload: Information and sensory overload in psychology and marketing. *Psychology and Marketing*, 1, 9–21

Manaker, G. H. (1996). *Interior plantscapes: Installation, maintenance, and management* (3rd edn). Englewood Cliffs, NJ: Prentice-Hall

Mancini, F. et al. (2011). Visual distortion of body size modulates pain perception. *Psychological Science*, 22, 325–30

Manning, J. T. and Fink, B. (2008). Digit ratio (2D:4D), dominance, reproductive success, asymmetry, and sociosexuality in the BBC Internet Study. *American Journal of Human Biology*, 20, 451–61

Manning, J. T. et al. (1998). The ratio of 2nd to 4th digit length: A predictor of sperm numbers and levels of testosterone, LH and oestrogen. *Human Reproduction*, 13, 3000–3004

Marin, M. M. et al. (2017). Misattribution of musical arousal increases sexual attraction towards opposite-sex faces in females. *PLOS One*, 12, e0183531

Marks, L. (1978). *The unity of the senses: Interrelations among the modalities.* New York: Academic Press

Martin, B. A. S. (2012). A stranger's touch: Effects of accidental interpersonal touch on consumer evaluations and shopping time. *Journal of Consumer Research*, 39, 174–84

Martin, S. (2013). How sensory information influences price decisions. *Harvard Business Review*, 26 July, https://hbr.org/2013/07/research-how-sensory-informati

Mathiesen, S. L. et al. (2020). Music to eat by: A systematic investigation of the relative importance of tempo and articulation on eating time. *Appetite*, 155, https://doi.org/10.1016/j.appet.2020.104801

Matsubayashi, T. et al. (2014). Does the installation of blue lights on train platforms shift suicide to another station? Evidence from Japan. *Journal of Affective Disorders*, 169, 57–60

Mattila, A. S. and Wirtz, J. (2001). Congruency of scent and music as a driver of in-store evaluations and behavior. *Journal of Retailing*, 77, 273–89

Mavrogianni, A. et al. (2013). Historic variations in winter indoor domestic temperatures and potential implications for body weight gain. *Indoor and Built Environment*, 22, 360–75

May, J. L. and Hamilton, P. A. (1980). Effects of musically evoked affect on women's interpersonal attraction toward and perceptual judgments of physical attractiveness of men. *Motivation and Emotion*, 4, 217–28

McCandless, C. (2011). *Feng shui that makes sense: Easy ways to create a home that feels as good as it looks.* Minneapolis, MN: Two Harbors Press

McCarty, K. et al. (2017). Optimal asymmetry and other motion parameters that characterise high-quality female dance. *Scientific Reports*, 7, 42435

McFarlane, S. J. et al. (2020). Alarm tones, music and their elements: A mixed methods analysis of reported waking sounds for the prevention of sleep inertia. *PLOS One*, 15, e0215788

McGann, J. P. (2017). Poor human olfaction is a 19th-century myth. *Science*, 356, eaam7263

McGlone, F. et al. (2013). The crossmodal influence of odor hedonics on facial attractiveness: Behavioral and fMRI measures. In F. Signorelli and D. Chirchiglia (eds.), *Functional Brain Mapping and the Endeavor to Understand the Working Brain*. Rijeka, Croatia: InTech Publications, pp. 209–25

McGuire, B. et al. (2018). Urine marking in male domestic dogs: Honest or dishonest? *Journal of Zoology*, 306, 163–70

McGurk, H. and MacDonald, J. (1976). Hearing lips and seeing voices. *Nature*, 264, 746–8

McKeown, J. D. and Isherwood, S. (2007). Mapping the urgency and pleasantness of speech, auditory icons, and abstract alarms to their referents within the vehicle. *Human Factors*, 49, 417–28

Mead, G. E. et al. (2009). Exercise for depression. *Cochrane Database Systematic Review*, CD004366

Mehta, R., Zhu, R. and Cheema, A. (2012). Is noise always bad? Exploring the effects of ambient noise on creative cognition. *Journal of Consumer Research*, 39, 784–99

Meijer, D. et al. (2019). Integration of audiovisual spatial signals is not consistent with maximum likelihood estimation. *Cortex*, 119, 74–88

Menzel, D. et al. (2008). Influence of vehicle color on loudness judgments. *Journal of the Acoustical Society of America*, 123, 2477–9

Merabet, L. B. et al. (2004). Visual hallucinations during prolonged blindfolding in sighted subjects. *Journal of Neuro-Ophthalmology*, 24, 109–13

Meston, C. M. and Frohlich, P. F. (2003). Love at first fright: Partner salience moderates roller-coaster-induced excitation transfer. *Archives of Sexual Behavior*, 32, 537–44

Meyers-Levy, J. and Zhu, R. (J.) (2007). The influence of ceiling height: The effect of priming on the type of processing that people use. *Journal of Consumer Research*, 34, 174–86

Mikellides, B. (1990). Color and physiological arousal. *Journal of Architectural and Planning Research*, 7, 13–20

Milgram, S. (1970). The experience of living in cities. *Science*, 167, 1461–8

Milinski, M. and Wedekind, C. (2001). Evidence for MHC-correlated perfume preferences in humans. *Behavioral Ecology*, 12, 140–49

Miller, G. et al. (2007). Ovulatory cycle effects on tip earnings by lap dancers: Economic evidence for human estrus? *Evolution and Human Behavior*, 28, 375–81

Miller, G. F. (1998). How mate choice shaped human nature: A review of sexual selection and human evolution. In C. B. Crawford and D. Krebs (eds.), *Handbook of evolutionary psychology: Ideas, issues, and applications*. Mahwah, NJ: Lawrence Erlbaum, pp. 87–129

——— (2000). Evolution of human music through sexual selection. In N. L. Wallin et al. (eds.), *The origins of music*. Cambridge, MA: MIT Press, pp. 329–60

Milliman, R. E. (1982). Using background music to affect the behavior of supermarket shoppers. *Journal of Marketing*, 46, 86–91

——— (1986). The influence of background music on the behavior of restaurant patrons. *Journal of Consumer Research*, 13, 286–9

Mindell, J. A. et al. (2009). A nightly bedtime routine: Impact on sleep in young children and maternal mood. *Sleep*, 32, 599–606

Minsky, L. et al. (2018). Inside the invisible but influential world of scent branding. *Harvard Business Review*, 11 April, https://hbr.org/2018/04/inside-the-invisible-but-influential-world-of-scent-branding

Mitchell, R. and Popham, F. (2008). Effect of exposure to natural environment on health inequalities: An observational population study. *The Lancet*, 372, 1655–60

Mitler, M. M. et al. (1988). Catastrophes, sleep, and public policy: Consensus report. *Sleep*, 11, 100–109

Mitro, S. et al. (2012). The smell of age: Perception and discrimination of body odors of different ages. *PLOS One*, 7, e38110

Miyazaki, Y. (2018). *Shinrin-yoku: The Japanese way of forest bathing for health and relaxation*. London: Aster Books

Monahan, J. L. et al. (2000). Subliminal mere exposure: Specific, general and affective effects. *Psychological Science*, 11, 462–6

Montagu, A. (1971). *Touching: The human significance of the skin*. New York: Columbia University Press

Montignies, F. et al. (2010). Empirical identification of perceptual criteria for customer-centred design. Focus on the sound of tapping on the dashboard when exploring a car. *International Journal of Industrial Ergonomics*, 40, 592–603

Moore, E. O. (1981). A prison environment's effect on health care service demands. *Journal of Environmental Systems*, 11, 17–34

Morgan, W. P. et al. (1988). Personality structure, mood states, and performance in elite male distance runners. *International Journal of Sport Psychology*, 19, 247–63

Morimoto, K. et al. (eds.) (2006). *Forest medicine*. Tokyo: Asakura Publishing

Morin, C. M. (1993). *Insomnia: Psychological assessment and management*. New York: Guilford Press

Morrin, M. and Chebat, J.-C. (2005). Person-place congruency: The interactive effects of shopper style and atmospherics on consumer expenditures. *Journal of Service Research*, 8, 181–91

Moseley, G. L. et al. (2008a). Is mirror therapy all it is cracked up to be? Current evidence and future directions. *Pain*, 138, 7–10

Moseley, G. L. et al. (2008b). Psychologically induced cooling of a specific body part caused by the illusory ownership of an artificial counterpart. *Proceedings of the National Academy of Sciences of the USA*, 105, 13168–72

Moseley, G. L. et al. (2008c). Visual distortion of a limb modulates the pain and swelling evoked by movement. *Current Biology*, 18, R1047–R1048

Moss, H. et al. (2007). A cure for the soul? The benefit of live music in the general hospital. *Irish Medical Journal*, 100, 636–8

Mueser, K. T. et al. (1984). You're only as pretty as you feel: Facial expression as a determinant of physical attractiveness. *Journal of Personality and Social Psychology*, 46, 469–78

Müller, F. et al. (2019). The sound of speed: How grunting affects opponents' anticipation in tennis. *PLOS One*, 14, e0214819

Mustafa, M. et al. (2016). The impact of vehicle fragrance on driving performance: What do we know? *Procedia – Social and Behavioral Sciences*, 222, 807–15

Muzet, A. et al. (1984). Ambient temperature and human sleep. *Experientia*, 40, 425–9

National Sleep Foundation (2006). *Teens and sleep*. https://sleepfoundation.org/sleep-topics/teens-and-sleep

Neave, N. et al. (2011). Male dance moves that catch a woman's eye. *Biology Letters*, 7, 221–4

Nettle, D. and Pollet, T. V. (2008). Natural selection on male wealth in humans. *American Naturalist*, 172, 658–66

Nieuwenhuis, M. et al. (2014). The relative benefits of green versus lean office space: Three field experiments. *Journal of Experimental Psychology: Applied*, 20, 199–214

Nightingale, F. (1860). *Notes on nursing. What it is, and what it is not.* New York: D. Appleton and Company

Nisbet, E. K. and Zelenski, J. M. (2011). Underestimating nearby nature: Affective forecasting errors obscure the happy path to sustainability. *Psychological Science*, 22, 1101–6

North, A. C. and Hargreaves, D. J. (1999). Music and driving game performance. *Scandinavian Journal of Psychology*, 40, 285–92

––––––– (2000). Musical preferences when relaxing and exercising. *American Journal of Psychology*, 113, 43–67

North, A. C., et al. (1997). In-store music affects product choice. *Nature*, 390, 132

North, A. C. et al. (1998). Musical tempo and time perception in a gymnasium. *Psychology of Music*, 26, 78–88

Novaco, R. et al. (1990). Objective and subjective dimensions of travel impedance as determinants of commuting stress. *American Journal of Community Psychology*, 18, 231–57

O'Connell, M. (2018). *To be a machine*. London: Granta

Oberfeld, D. et al. (2009). Ambient lighting modifies the flavor of wine. *Journal of Sensory Studies*, 24, 797–832

Oberfeld, D. et al. (2010). Surface lightness influences perceived room height. *Quarterly Journal of Experimental Psychology*, 63, 1999–2011

Obst, P. et al. (2011). Age and gender comparisons of driving while sleepy: Behaviours and risk perceptions. *Transportation Research Part F: Traffic Psychology and Behaviour*, 14, 539–42

Oldham, G. R. et al. (1995). Listen while you work? Quasi-experimental relations between personal-stereo headset use and employee work responses. *Journal of Applied Psychology*, 80, 547–64

Olmsted, F. L. (1865a). The value and care of parks. Reprinted in R. Nash (ed.) (1968). *The American environment: Readings in the history of conservation*. Reading, MA: Addison-Wesley, pp. 18–24

––––––– (1865b). *Yosemite and the Mariposa Grove: A preliminary report*. Available online at: www.yosemite.ca.us/library/olmsted/report.html

Olson, R. L. et al. (2009). Driver distraction in commercial vehicle operations. Technical Report No. FMCSA-RRR-09-042. Federal Motor Carrier Safety Administration, US Department of Transportation, Washington, DC

Olsson, M. J. et al. (2014). The scent of disease: Human body odor contains an early chemosensory cue of sickness. *Psychological Science*, 25, 817–23

Ott, W. R. and Roberts, J. W. (1998). Everyday exposure to toxic pollutants. *Scientific American*, 278 (February), 86–91

Otterbring, T. (2018). Healthy or wealthy? Attractive individuals induce sex-specific food preferences. *Food Quality and Preference*, 70, 11–20

Otterbring, T. et al. (2018). The relationship between office type and job satisfaction: Testing a multiple mediation model through ease of interaction and well-being. *Scandinavian Journal of Work and Environmental Health*, 44, 330–34

Ottoson, J. and Grahn, P. (2005). A comparison of leisure time spent in a garden with leisure time spent indoors: On measures of restoration in residents in geriatric care. *Landscape Research*, 30, 23–55

Oyer, J. and Hardick, J. (1963). *Response of population to optimum warning signal*. Office of Civil Defence, Final Report No. SHSLR163. Contract No. OCK-OS-62-182, September

Packard, V. (1957). *The hidden persuaders*. Harmondsworth: Penguin

Pallasmaa, J. (1996). *The eyes of the skin: Architecture and the senses (Polemics)*. London: Academy Editions

Palmer, H. (1978). *Sea gulls . . . Music for rest and relaxation*. Freeport, NY: Education Activities, Inc. (Tape #AR504)

Pancoast, S. (1877). *Blue and red light*. Philadelphia: J. M. Stoddart & Co.

Park, B. J. et al. (2007). Physiological effects of Shinrin-yoku (taking in the atmosphere of the forest) – using salivary cortisol and cerebral activity as indicators. *Journal of Physiological Anthropology*, 26, 123–8

Park, J. and Hadi, R. (2020). Shivering for status: When cold temperatures increase product evaluation. *Journal of Consumer Psychology*, 30, 314–28

Park, Y.-M. M. et al. (2019). Association of exposure to artificial light at night while sleeping with risk of obesity in women. *JAMA Internal Medicine*, 179, 1061–71

Parsons, R. et al. (1998). The view from the road: Implications for stress recovery and immunization. *Journal of Environmental Psychology*, 18, 113–40

Passchier-Vermeer, W. and Passchier, W. F. (2000). Noise exposure and public health. *Environmental Health Perspectives*, 108, 123–31

Pasut, W. et al. (2015). Energy-efficient comfort with a heated/cooled chair: Results from human subject tests. *Building and Environment*, 84, 10–21

Patania, V. M. et al. (2020). The psychophysiological effects of different tempo music on endurance versus high-intensity performances. *Frontiers in Psychology*, 11, 74

Pavela Banai, I. (2017). Voice in different phases of menstrual cycle among naturally cycling women and users of hormonal contraceptives. *PLOS One*, 12, e0183462

Peck, J. and Shu, S. B. (2009). The effect of mere touch on perceived ownership. *Journal of Consumer Research*, 36, 434–47

Peltzman, S. (1975). The effects of automobile safety regulation. *Journal of Political Economy*, 83, 677–725

Pencavel, J. (2014). The productivity of working hours. IZA Discussion Paper No. 8129, http://ftp.iza.org/dp8129.pdf

Peperkoorn, L. S. et al. (2016). Revisiting the red effect on attractiveness and sexual receptivity: No effect of the color red on human mate preference. *Evolutionary Psychology*, October–December, 1–13

Perrault, A. A. et al. (2019). Whole-night continuous rocking entrains spontaneous neural oscillations with benefits for sleep and memory. *Current Biology*, 29, 402–11

Petit, O. et al. (2019). Multisensory consumer-packaging interaction (CPI): The role of new technologies. In C. Velasco and C. Spence (eds.), *Multisensory packaging: Designing new product experiences*. Cham, Switzerland: Palgrave Macmillan, pp. 349–74

Pfaffmann, C. (1960). The pleasure of sensation. *Psychological Review*, 67, 253–68

Phalen, J. M. (1910). An experiment with orange-red underwear. *Philippine Journal of Science*, 5B, 525–46

Pinker, S. (2018). *Enlightenment now: The case for reason, science, humanism, and progress*. New York: Viking Penguin

Piqueras-Fiszman, B. and Spence, C. (2012). The weight of the bottle as a possible extrinsic cue with which to estimate the price (and quality) of the wine? Observed correlations. *Food Quality and Preference*, 25, 41–5

Plante, T. G. et al. (2006). Psychological benefits of exercise paired with virtual reality: Outdoor exercise energizes whereas indoor virtual exercise relaxes. *International Journal of Stress Management*, 13, 108–17

Pollet, T. et al. (2018). Do red objects enhance sexual attractiveness? No evidence from two large replications and an extension. PsyArXiv Preprints, 16 February 2018, https://doi.org/10.31234/osf.io/3bfwh

Prescott, J. and Wilkie, J. (2007). Pain tolerance selectively increased by a sweet-smelling odor. *Psychological Science*, 18, 308–11

Pretty, J. et al. (2009). *Nature, childhood, health and life pathways*. University of Essex, Interdisciplinary Centre for Environment and Society, Occasional Paper 2009-2

Priest, D. L. et al. (2004). The characteristics and effects of motivational music in exercise settings: The possible influence of gender, age, frequency of attendance, and time of attendance. *Journal of Sports Medicine and Physical Fitness*, 44, 77–86

Przybylski, A. K. (2019). Digital screen time and pediatric sleep: Evidence from a preregistered cohort study. *Journal of Pediatrics*, 205, 218–23

Qin, J. et al. (2014). The effect of indoor plants on human comfort. *Indoor Building Environment*, 23, 709–23

Ramachandran, V. S. and Blakeslee, S. (1998). *Phantoms in the brain*. London: Fourth Estate

Ramsey, K. L. and Simmons, F. B. (1993). High-powered automobile stereos. *Otolaryngology – Head and Neck Surgery*, 109, 108–10

Ratcliffe, E. et al. (2016). Associations with bird sounds: How do they relate to perceived restorative potential? *Journal of Environmental Psychology*, 47, 136–44

Ratcliffe, V. F. et al. (2016). Cross-modal correspondences in non-human mammal communication. *Multisensory Research*, 29, 49–91

Rattenborg, N. C. et al. (1999). Half-awake to the risk of predation. *Nature*, 397, 397–8

Raudenbush, B. et al. (2001). Enhancing athletic performance through the administration of peppermint odor. *Journal of Sport and Exercise Psychology*, 23, 156–60

Raudenbush, B. et al. (2002). The effects of odors on objective and subjective measures of athletic performance. *International Sports Journal*, 6, 14–27

Raymann, R. J. et al. (2008). Skin deep: Enhanced sleep depth by cutaneous temperature manipulation. *Brain*, 131, 500–513

Raymond, J. (2000). The world of senses. *Newsweek Special Issue*, Fall–Winter, 136, 16–18

Reber, R., et al. (2004). Processing fluency and aesthetic pleasure: Is beauty in the perceiver's processing experience? *Personality and Social Psychology Review*, 8, 364–82

Reber, R., et al. (1998). Effects of perceptual fluency on affective judgments. *Psychological Science*, 9, 45–8

Redelmeier, D. A. and Tibshirani, R. J. (1997). Association between cellular-telephone calls and motor vehicle collisions. *New England Journal of Medicine*, 336, 453–8

Redies, C. (2007). A universal model of esthetic perception based on the sensory coding of natural stimuli. *Spatial Vision*, 21, 97–117

Reinoso-Carvalho, F. et al. (2019). Not just another pint! Measuring the influence of the emotion induced by music on the consumer's tasting experience. *Multisensory Research*, 32, 367–400

Renvoisé, P. and Morin, C. (2007). *Neuromarketing: Understanding the 'buy buttons' in your customer's brain*. Nashville, TN: Thomas Nelson

Rhodes, G. (2006). The evolutionary psychology of facial beauty. *Annual Review of Psychology*, 57, 199–226

Rice, T. (2003). Soundselves: An acoustemology of sound and self in the Edinburgh Royal Infirmary. *Anthropology Today*, 19, 4–9

Richter, J. and Muhlestein, D. (2017). Patient experience and hospital profitability: Is there a link? *Health Care Management Review*, 42, 247–57

Roberts, S. C. et al. (2004). Female facial attractiveness increases during the fertile phase of the menstrual cycle. *Proceedings of the Royal Society of London Series B*, 271 (S5), S270–S272

Roberts, S. C. et al. (2011). Body odor quality predicts behavioral attractiveness in humans. *Archives of Sexual Behavior*, 40, 1111–17

Roenneberg, T. (2012). *Internal time: Chronotypes, social jet lag, and why you're so tired*. Cambridge, MA: Harvard University Press

——— (2013). Chronobiology: The human sleep project. *Nature*, 498, 427–8

Romero, J. et al. (2003). Color coordinates of objects with daylight changes. *Color Research and Application*, 28, 25–35

Romine, I. J. et al. (1999). Lavender aromatherapy in recovery from exercise. *Perceptual and Motor Skills*, 88, 756–8

Roschk, H. et al. (2017). Calibrating 30 years of experimental research: A meta-analysis of the atmospheric effects of music, scent, and color. *Journal of Retailing*, 93, 228–40

Rosenblum, L. D. (2010). *See what I am saying: The extraordinary powers of our five senses*. New York: W. W. Norton

Rosenthal, N. E (2019). *Winter blues: Everything you need to know to beat Seasonal Affective Disorder*. New York: Guilford Press

Ross, S. (1966). Background music systems – do they pay? *Administrative Management Journal*, 27 (August), 34–7

Rowe, C. et al. (2005). Seeing red? Putting sportswear in context. *Nature*, 437, E10

Rybkin, I. (2017). Music's potential effects on surgical performance. *Quill and Scope*, 10, 3

Sagberg, F. (1999). Road accidents caused by drivers falling asleep. *Accident Analysis and Prevention*, 31, 639–49

Salgado-Montejo., A. et al. (2015). Smiles over frowns: When curved lines influence product preference. *Psychology and Marketing*, 32, 771–81

Samuel, L. R. (2010). *Freud on Madison Avenue: Motivation research and subliminal advertising in America*. Oxford: University of Pennsylvania Press

Schaal, B. and Durand, K. (2012). The role of olfaction in human multisensory development. In A. J. Bremner et al. (eds.), *Multisensory development*. Oxford: Oxford University Press, pp. 29–62

Schaal, B. et al. (2000). Human foetuses learn odours from their pregnant mother's diet. *Chemical Senses*, 25, 729–37

Schaefer, E. W. et al. (2012). Sleep and circadian misalignment for the hospitalist: A review. *Journal of Hospital Medicine*, 7, 489–96

Schaffert, N. et al. (2011). An investigation of online acoustic information for elite rowers in on-water training conditions. *Journal of Human Sport and Exercise*, 6, 392–405

Schiffman, S. S. and Siebert, J. M. (1991). New frontiers in fragrance use. *Cosmetics and Toiletries*, 106, 39–45

Scholey, A. et al. (2009). Chewing gum alleviates negative mood and reduces cortisol during acute laboratory psychological stress. *Physiology and Behavior*, 97, 304–12

Schreiner, T. and Rasch, B. (2015). Boosting vocabulary learning by verbal cueing during sleep. *Cerebral Cortex*, 25, 4169–79

Schreuder, E. et al. (2016). Emotional responses to multisensory environmental stimuli: A conceptual framework and literature review. *Sage Open*, January–March, 1–19

Schwartzman, M. (2011). *See yourself sensing: Redefining human perception*. London: Black Dog

Sekuler, R. and Blake, R. (1987). Sensory underload. *Psychology Today*, 12 (December), 48–51

Seligman, M. E. (1971). Phobias and preparedness. *Behavior Therapy*, 2, 307–20

Senders, J. W. et al. (1967). The attentional demand of automobile driving. *Highway Research Record*, 195, 15–33

Senkowski, D. et al. (2014). Crossmodal shaping of pain: A multisensory approach to nociception. *Trends in Cognitive Sciences*, 18, 319–27

Seto, K. C. et al. (2012). Global forecasts of urban expansion to 2030 and direct impacts on biodiversity and carbon pools. *Proceedings of the National Academy of Sciences of the USA*, 109, 16083–8

Sheldon, R. and Arens, E. (1932). *Consumer engineering: A new technique for prosperity*. New York: Harper & Brothers

Shippert, R. D. (2005). A study of time-dependent operating room fees and how to save $100 000 by using time-saving products. *American Journal of Cosmetic Surgery*, 22, 25–34

Sinnett, S. and Kingstone, A. (2010). A preliminary investigation regarding the effect of tennis grunting: Does white noise during a tennis shot have a negative impact on shot perception? *PLOS One*, 5, e13148

Sitwell, W. (2020). *The restaurant: A history of eating out*. London: Simon & Schuster

Sivak, M. (1996). The information that drivers use: Is it indeed 90% visual? *Perception*, 25, 1081–9

Siverdeen, Z. et al. (2008). Exposure to noise in orthopaedic theatres – do we need protection? *International Journal of Clinical Practice*, 62, 1720–22

Slabbekoorn, H. and Ripmeester, E. (2008). Birdsong and anthropogenic noise: Implications and applications for conservation. *Molecular Ecology*, 17, 72–83

Smith, G. A. et al. (2006). Comparison of a personalized parent voice smoke alarm with a conventional residential tone smoke alarm for awakening children. *Pediatrics*, 118, 1623–32

Smith, M. M. (2007). *Sensory history*. Oxford: Berg

Solomon, M. R. (2002). *Consumer behavior: Buying, having and being*. Upper Saddle River, NJ: Prentice-Hall

Sorokowska, A. et al. (2012). Does personality smell? Accuracy of personality assessments based on body odour. *European Journal of Personality*, 26, 496–503

Sors, F. et al. (2017). The contribution of early auditory and visual information to the discrimination of shot power in ball sports. *Psychology of Sport and Exercise*, 31, 44–51

Souman, J. L. et al. (2017). Acute alerting effects of light: A systematic literature review. *Behavioural Brain Research*, 337, 228–39

Spence, C. (2002). *The ICI report on the secret of the senses*. London: The Communication Group

——— (2003). A new multisensory approach to health and well-being. *In Essence*, 2, 16–22

——— (2012a). Drive safely with neuroergonomics. *The Psychologist*, 25, 664–7

——— (2012b). Managing sensory expectations concerning products and brands: Capitalizing on the potential of sound and shape symbolism. *Journal of Consumer Psychology*, 22, 37–54

——— (2014). Q & A: Charles Spence. *Current Biology*, 24, R506–R508

——— (2015). Leading the consumer by the nose: On the commercialization of olfactory-design for the food and beverage sector. *Flavour*, 4, 31

——— (2016). Gastrodiplomacy: Assessing the role of food in decision-making. *Flavour*, 5, 4

——— (2017). Hospital food. *Flavour*, 6, 3

——— (2018). *Gastrophysics: The new science of eating*. London: Penguin

——— (2019a). Attending to the chemical senses. *Multisensory Research*, 32, 635–64

——— (2019b). Multisensory experiential wine marketing. *Food Quality and Preference*, 71, 106–16, https://doi.org/10.1016/j.foodqual.2018.06.010

——— (2020a). Extraordinary emotional responses elicited by auditory stimuli linked to the consumption of food and drink. *Acoustical Science and Technology*, 41, 28–36

——— (2020b). Multisensory flavour perception: Blending, mixing, fusion, and pairing within and between the senses. *Foods*, 9, 407

——— (2020c). On the ethics of neuromarketing and sensory marketing. In J. Trempe-Martineau and E. Racine (eds.), *Organizational neuroethics: Reflections on the contributions of neuroscience to management theories and business practice*. Cham, Switzerland: Springer Nature, pp. 9–30

——— (2020d). Temperature-based crossmodal correspondences: Causes and consequences. *Multisensory Research*, 33, 645–82

——— (2020e). Designing for the multisensory mind. *Architectural Design*, December, 42-49

——— (2020f). Senses of space: Designing for the multisensory mind. *Cognitive Research: Principles and Implications*, 5, 46. https://rdcu.be/b7qIt

Spence, C. and Carvalho, F. M. (2020). The coffee drinking experience: Product extrinsic (atmospheric) influences on taste and choice. *Food Quality and Preference*, 80, https://doi.org/10.1016/j.foodqual.2019.103802

Spence, C. and Gallace, A. (2011). Multisensory design: Reaching out to touch the consumer. *Psychology and Marketing*, 28, 267–308

Spence, C. and Keller, S. (2019). Medicine's melodies: On the costs and benefits of music, soundscapes, and noise in healthcare settings. *Music and Medicine*, 11, 211–25

Spence, C. and Read, L. (2003). Speech shadowing while driving: On the difficulty of splitting attention between eye and ear. *Psychological Science*, 14, 251–6

Spence, C. et al. (2014a). A large sample study on the influence of the multisensory environment on the wine drinking experience. *Flavour*, 3, 8

Spence, C. et al. (2014b). Store atmospherics: A multisensory perspective. *Psychology and Marketing*, 31, 472–88

Spence, C. et al. (2017). Digitizing the chemical senses: Possibilities and pitfalls. *International Journal of Human-Computer Studies*, 107, 62–74

Spence, C. et al. (2019a). Digital commensality: On the pros and cons of eating and drinking with technology. *Frontiers in Psychology*, 10, 2252

Spence, C. et al. (2019b). Extrinsic auditory contributions to food perception and consumer behaviour: An interdisciplinary review. *Multisensory Research*, 32, 275–318

Spence, C. et al. (2020). Magic on the menu: Where are all the magical food and beverage experiences? *Foods*, 9, 257

Stack, S. and Gundlach, J. (1992). The effect of country music on suicide. *Social Forces*, 71, 211–18

Stanton, T. R. et al. (2017). Feeling stiffness in the back: A protective perceptual inference in chronic back pain. *Scientific Reports*, 7, 9681

Staricoff, R. and Loppert, S. (2003). Integrating the arts into health care: Can we affect clinical outcomes? In D. Kirklin and R. Richardson (eds.), *The healing environment: Without and within*. London: RCP, pp. 63–79

Steel, C. (2008). *Hungry city: How food shapes our lives*. London: Chatto & Windus

Steele, K. M. (2014). Failure to replicate the Mehta and Zhu (2009) color-priming effect on anagram solution times. *Psychonomic Bulletin and Review*, 21, 771–6

Stein, B. E. (ed.-in-chief) (2012). *The new handbook of multisensory processing*. Cambridge, MA: MIT Press

Steinwald, M. et al. (2014). Multisensory engagement with real nature relevant to real life. In N. Levent and A. Pascual-Leone (eds.), *The multisensory museum: Cross-disciplinary perspectives on touch, sound, smell, memory and space*. Plymouth: Rowman & Littlefield, pp. 45–60

Stillman, J. W. and Hensley, W. E. (1980). She wore a flower in her hair: The effect of ornamentation on non-verbal communication. *Journal of Applied Communication Research*, 1, 31–9

Stumbrys, T. et al. (2012). Induction of lucid dreams: A systematic review of evidence. *Consciousness and Cognition*, 21, 1456–75

Suwabe, K. et al. (in press). Positive mood while exercising influences beneficial effects of exercise with music on prefrontal executive function: A functional NIRS Study. *Neuroscience*, https://doi.org/10.1016/j.neuroscience.2020.06.007

Taheri, S. et al. (2004). Short sleep duration is associated with reduced leptin, elevated ghrelin, and increased body mass index. *PLOS Medicine*, 1, 210–17

Tamaki, M. et al. (2016). Night watch in one brain hemisphere during sleep associated with the first-night effect in humans. *Current Biology*, 26, 1190–94

Tanizaki, J. (2001). *In praise of shadows* (trans. T. J. Harper and E. G. Seidenstickker). London: Vintage Books

Tassi, P. and Muzet, A. (2000). Sleep inertia. *Sleep Medicine Reviews*, 4, 341–53

Terman, M. (1989). On the question of mechanism in phototherapy for seasonal affective disorder: Considerations of clinical efficacy and epidemiology. In N. E. Rosenthal and M. C. Blehar (eds.), *Seasonal affective disorders and phototherapy*. New York: Guilford Press, pp. 357–76

Terry, P. C. et al. (2012). Effects of synchronous music on treadmill running among elite triathletes. *Journal of Science and Medicine in Sport*, 15, 52–7

Thömmes, K. and Hübner, R. (2018). Instagram likes for architectural photos can be predicted by quantitative balance measures and curvature. *Frontiers in Psychology*, 9, 1050

Thompson Coon, J. et al. (2011). Does participating in physical activity in outdoor natural environments have a greater effect on physical and mental wellbeing than physical activity indoors? A systematic review. *Environmental Science and Technology*, 45, 1761–72

Tifferet, S. et al. (2012). Guitar increases male Facebook attractiveness: Preliminary support for the sexual selection theory of music. *Letters on Evolutionary Behavioral Science*, 3, 4–6

Townsend, M. and Weerasuriya, R. (2010). *Beyond blue to green: The benefits of contact with nature for mental health and well-being.* Melbourne, Australia: Beyond Blue Limited

Treib, M. (1995). Must landscape mean? Approaches to significance in recent landscape architecture. *Landscape Journal*, 14, 47–62

Treisman, M. (1977). Motion sickness: As evolutionary hypothesis. *Science*, 197, 493–5

Trivedi, B. (2006). Recruiting smell for the hard sell. *New Scientist*, 2582, 36–9

Trotti, L. M. (2017). Waking up is the hardest thing I do all day: Sleep inertia and sleep drunkenness. *Sleep Medicine Reviews*, 35, 76–84

Trzeciak, S. et al. (2016). Association between Medicare summary star ratings for patient experience and clinical outcomes in US hospitals. *Journal of Patient Experience*, 3, 6–9

Tse, M. M. et al. (2002). The effect of visual stimuli on pain threshold and tolerance. *Journal of Clinical Nursing*, 11, 462–9

Twedt, E. et al. (2016). Designed natural spaces: Informal gardens are perceived to be more restorative than formal gardens. *Frontiers in Psychology*, 7, 88

Ullmann, Y. et al. (2008). The sounds of music in the operating room. *Injury*, 39, 592–7

Ulrich, R. S. (1984). View through a window may influence recovery from surgery. *Science*, 224, 420–21

——— (1991). Effects of interior design on wellness: Theory and recent scientific research. *Journal of Health Care Interior Design*, 3, 97–109

——— (1993). Biophilia, biophobia, and natural landscapes. In S. R. Kellert and E. O. Wilson (eds.), *The biophilia hypothesis*. Washington, DC: Island Press, pp. 73–137

——— (1999). Effects of gardens on health outcomes: Theory and research. In C. Cooper-Marcus and M. Barnes (eds.), *Healing gardens: Therapeutic benefits and design recommendations*. Hoboken, NJ: John Wiley & Sons, pp. 27–86

Ulrich, R. S. et al. (1991). Stress recovery during exposure to natural and urban environments. *Journal of Environmental Psychology*, 11, 201–30

Underhill, P. (1999). *Why we buy: The science of shopping*. New York: Simon & Schuster

Unkelbach, C. and Memmert, D. (2010). Crowd noise as a cue in referee decisions contributes to the home advantage. *Journal of Sport and Exercise Psychology*, 32, 483–98

Unnava, V. et al. (2018). Coffee with co-workers: Role of caffeine on evaluations of the self and others in group settings. *Journal of Psychopharmacology*, 32, 943–8

Ury, H. K. et al. (1972). Motor vehicle accidents and vehicular pollution in Los Angeles. *Archives of Environmental Health*, 25, 314–22

US Energy Information Administration (2011). Residential energy consumption survey (RECS). *US Energy Information Administration*, www.eia.gov/consumption/residential/reports/2009/air-conditioning.php

US Senate Special Committee on Aging (1985–6). *Aging America, Trends and Projections, 1985–86 Edition*. US Senate Special Committee on Aging (in association with the American Association of Retired Persons, the Federal Council on the Aging, and the Administration on Aging)

Valdez, P. and Mehrabian, A. (1994). Effects of color on emotions. *Journal of Experimental Psychology: General*, 123, 394–409

Vartanian, O. et al. (2013). Impact of contour on aesthetic judgments and approach-avoidance decisions in architecture. *Proceedings of the National Academy of Sciences of the USA*, 110 (Supplement 2), 10446–53

Vartanian, O. et al. (2015). Architectural design and the brain: Effects of ceiling height and perceived enclosure on beauty judgments and approach-avoidance decisions. *Journal of Environmental Psychology*, 41, 10–18

Villemure, C. et al. (2003). Effects of odors on pain perception: Deciphering the roles of emotion and attention. *Pain*, 106, 101–8

Wagner, U. et al. (2004). Sleep inspires insight. *Nature*, 427, 352–5

Walker, J. et al. (2016). Chewing unflavored gum does not reduce cortisol levels during a cognitive task but increases the response of the sympathetic nervous system. *Physiology and Behavior*, 154, 8–14

Walker, M. (2018). *Why we sleep*. London: Penguin

Wallace, A. G. (2015). Are you looking at me? *Capital Ideas*, Fall, 24–33

Wang, Q. J. and Spence, C. (2019). Drinking through rosé-coloured glasses: Influence of wine colour on the perception of aroma and flavour in wine experts and novices. *Food Research International*, 126, 108678

Wargocki, P. et al. (1999). Perceived air quality, Sick Building Syndrome (SBS) symptoms and productivity in an office with two different pollution loads. *Industrial Air*, 9, 165–79

Wargocki, P. et al. (2000). The effects of outdoor air supply rate in an office on perceived air quality, Sick Building Syndrome (SBS) symptoms and productivity. *Industrial Air*, 10, 222–36

Warm, J. S. et al. (1991). Effects of olfactory stimulation on performance and stress in a visual sustained attention task. *Journal of the Society of Cosmetic Chemists*, 42, 199–210

Waterhouse, J. et al. (2010). Effects of music tempo upon submaximal cycling performance. *Scandinavian Journal of Medicine and Science in Sports*, 20, 662–9

Watkins, C. D. (2017). Creativity compensates for low physical attractiveness when individuals assess the attractiveness of social and romantic partners. *Royal Society Open Science*, 4, 160955

Watson, L. (1971). *The omnivorous ape.* New York: Coward, McCann & Geoghegan

Weber, S. T. and Heuberger, E. (2008). The impact of natural odors on affective states in humans. *Chemical Senses,* 33, 441–7

Wehrens, S. M. T. et al. (2017). Meal timing regulates the human circadian system. *Current Biology,* 27, 1768–75

Weinzimmer, D. et al. (2014). Human responses to simulated motorized noise in national parks. *Leisure Sciences,* 36, 251–67

Whalen, P. J. et al. (2004). Human amygdala responsivity to masked fearful eye whites. *Science,* 306, 2061

Whitcome, K. K. et al. (2007). Fetal load and the evolution of lumbar lordosis in bipedal hominins. *Nature,* 450, 1075–8

White, D. et al. (2017). Choosing face: The curse of self in profile image selection. *Cognitive Research: Principles and Implications,* 2, 23

Wigley, M. (1995). *White walls, designer dresses: The fashioning of modern architecture.* London: MIT Press

Wilde, G. J. S. (1982). The theory of risk homeostasis: Implications for safety and health. *Risk Analysis,* 2, 209–25

Williams, F. (2017). *The nature fix: Why nature makes us happier, healthier, and more creative.* London: W. W. Norton & Co.

Willis, J. and Todorov, A. (2006). First impressions: Making up your mind after a 100-ms exposure to a face. *Psychological Science,* 17, 592–8

Wilson, E. O. (1984). *Biophilia: The human bond with other species.* London: Harvard University Press

Wilson, T. D. and Gilbert, D. T. (2005). Affective forecasting: Knowing what to want. *Current Directions in Psychological Science,* 14, 131–4

Windhager, S. et al. (2008). Face to face: The perception of automotive designs. *Human Nature,* 19, 331–46

Winternitz, J. et al. (2017). Patterns of MHC-dependent mate selection in humans and nonhuman primates: A meta-analysis. *Molecular Ecology,* 26, 668–88

Wittkopf, P. G. et al. (2018). The effect of visual feedback of body parts on pain perception: A systematic review of clinical and experimental studies. *European Journal of Pain*, 22, 647–62

Wohlwill, J. F. (1983). The concept of nature: A psychologist's view. In I. Altman and J. F. Wohlwill (eds.), *Behavior and the natural environment*. New York: Plenum Press, pp. 5–38

Wolverton, B. C. et al. (1989). *Interior landscape plants for indoor air pollution abatement*. Final Report, 15 September. National Aeronautics and Space Administration, John C. Stennis Space Center, Science and Technology Laboratory, Stennis Space Center, MS 39529–6000

Wood, R. A. et al. (2006). The potted-plant microcosm substantially reduces indoor air VOC pollution: I. Office field-study. *Water, Air, and Soil Pollution*, 175, 163–80

Woolley, K. and Fishbach, A. (2017). A recipe for friendship: Similar food consumption promotes trust and cooperation. *Journal of Consumer Psychology*, 27, 1–10

World Health Organization, Regional Office for Europe (2011). *Burden of disease from environmental noise – Quantification of healthy life years lost in Europe*. Copenhagen: WHO

Wright, K. P., Jr and Czeisler, C. A. (2002). Absence of circadian phase resetting in response to bright light behind the knees. *Science*, 297, 571

Wrisberg, C. A. and Anshel, M. H. (1989). The effect of cognitive strategies on free throw shooting performance of young athletes. *Sport Psychologist*, 3, 95–104

Yildirim, K. et al. (2007). The effects of window proximity, partition height, and gender on perceptions of open-plan offices. *Journal of Environmental Psychology*, 27, 154–65

Yoder, J. et al. (2012). Noise and sleep among adult medical inpatients: Far from a quiet night. *Archives of Internal Medicine*, 172, 68–70

Zellner, D. et al. (2017). Ethnic congruence of music and food affects food selection but not liking. *Food Quality and Preference*, 56, 126–9

Zhang, Y. et al. (2019). Healing built-environment effects on health outcomes: Environment–occupant–health framework. *Building Research and Information*, 47, 747–66

Zhu, R. (J.) and Argo, J. J. (2013). Exploring the impact of various shaped seating arrangements on persuasion. *Journal of Consumer Research*, 40, 336–49

Ziegler, U. (2015). Multi-sensory design as a health resource: Customizable, individualized, and stress-regulating spaces. *Design Issues*, 31, 53–62

Zilczer, J. (1987). 'Color music': Synaesthesia and nineteenth-century sources for abstract art. *Artibus et Historiae*, 8, 101–26

Zuckerman, M. (1979). *Sensation seeking: Beyond the optimal level of arousal.* Hillsdale, NJ: Lawrence Erlbaum

# Index

Page references in *italic* indicate illustrations or their captions.

2D:4D digit ratio 230
7UP logo 143–4, *143*, 155
23andMe 77, 78

Abercrombie & Fitch 158, 162, 167
Accenture 1
Ackerman, Diane: *A natural history of the senses* 61
Agassi, Andre 210
'ah-ha' moments 74
air conditioning 121–3
air fresheners 79, 80
Air India Express crash (2010) 88
air quality, indoor 77, 120, 133–4
alarm clocks 85
Alberti, Leon 107
All Blacks 217
Allen, K. 175
aloe vera 76–7
Alzheimer's 181–2, 195
Amazon 9, *10*
  Seattle offices 134–5, *135*
Amherst, Alicia: *A history of gardening in England* 58
amygdala 49
'ancestral health' movement 203–4
animal therapy 195
Anne Fontaine stores 157
anthropomorphism, in product design *8*, 9–10
Antonioni, Michelangelo 28

Antonovsky, Aaron 179
aphrodisiacs 240–41
appetite suppressants 31–4
Apple 23, 139
Appleyard, D. 108n
Arens, E. 98–9
Argos 9
Aristotle 90
aromatherapy 20, 52, 84, 212
  essential oils 23n, 52, 84, 111, 212
  massage 195, 255
art, use in healthcare 183–5
Asahi 3
Asda 161
Ashley Madison dating site 239
associative learning 20, 94
AT&T, 'It can wait' campaign 103
atmospherics in shopping 156–69
  overload/incongruence dangers 165–8
  scents 157–9
  and superadditivity 164–7
  and tactile contamination 163–4
  temperature 160
  and touch 161–4
attention restoration theory (ART) 48–9
autonomous sensory meridian response (ASMR) content 74, 171–2
avoidance motivation 218–19

Babbitt, Edwin 184
babies
  foetal soap syndrome 7
  need of balanced mix of multisensory stimulation 3
  turning towards aromas of foods mothers ate during pregnancy 7
Babitz, Liviu 260–61
Bailly Dunne, Catherine 25
Baker-Miller pink 32–3, 184
Bakker, Michael 139
Balachandra, Lakshmi 140
Bang & Olufsen 162–3
Barnes & Noble 147
Barton, R. A. 219
bathing 38–41
  scented baths 81
  taking a bath before sleeping 75, 81, 265
  *see also* forest bathing
bathrooms 26
  showering or taking a bath 38–41, 75, 81, 265
Baudelaire, Charles 18
beauty 222, 236, 238
  as big business 222
  of nature on a computer screen 135–7
  and the voice 242–3
  *see also* sexual attraction and dating
'beer goggles' effect 241–2
Bentley Continental GT 98
benzodiazepine-receptor agonists 68
Berg, A. van den 62
big data analysis 144
biodiversity 51–2
biophilia hypothesis (Wilson) 44–5, 47, 49, 52, 60–61
biophobia 61
birdsong 51–2, 131, 265
  dawn chorus 57, 85
Blascovich, J. 175

*Blue Wings* 45
Blumenthal, Heston, *Sound of the Sea* dish 182
bodily rhythms *see* circadian clock/rhythms
body odour 230–32, 245
body posture 217
body temperature 75–6
Botox 223n
boxing 212, 217
brain 49–50, 52–3, 73, 93, 101
  amygdala 49
  central governor hypothesis 214
  and driving 103–4, 113, 115
  and exercise/sport 199, 209–10, 214
  and eye contact 228
  hemispheres 79–80
  Neuralink and chips for insertion into 262
  neuroimaging 228
  neuroscience *see* consumer neuroscience; neuroscience
  orbitofrontal cortex 234
  predictive coding 214
  and red colour 219
  and sense of smell 80
  shoppers' brains 144 *see also* consumer neuroscience
  ventral striatum 228
bread smells 144–5, 266
Brick, Neil 213
Brighton, Prince Regent's Royal Pavilion 24, *24*
'Britain's Healthiest Workplace' 67
*British Medical Journal* 174
building odour 18–20, 21
*Businessweek* 150

## Index

caffeine *see* coffee/caffeine
Campos, Claudia 181
Cannes Lions festival 147–8
car sickness 93, 112–14
carbohydrate drinks/rinsing 213–14, 266
carbon dioxide 120, 134
Carême, Marie Antoine (Antonin) 24
cars 92–115, 265
   alerts for drowsy drivers 104–6, 110
   colour 99
   commuting by car *see* drivers and commuting by car
   door and interior sounds 97–8
   electric/hybrid 92, 97, 115
   engine sound 95–7
   feel of 98–9
   keys 98
   and marketers 92
   multisensory design and hacking of drivers' senses 93–9
   'new-car-smell' 93–5
   scent displays in 109–11
   semi-autonomous 92, 115
   technology 92–3
Casanova, Giacomo 241
ceiling height 23–4
Centers for Disease Control and Prevention (CDC) 66
central governor hypothesis 214
*Challenger* Space Shuttle disaster 88
Chanel No. 5 157
Chebat, J. C. 165–6, 168
Chekroud, Adam 200
Chernobyl nuclear disaster 88
Cheskin, Louis 142–3, 155, 163
chewing gum 214
Chipotle 150–51
Christian Louboutin brand 238
chromatherapy 184, 198

chronic regional pain syndrome (CRPS) 185, 187–8
Churchill, Winston 39
CIA 248–9
Cinnabon 145
circadian clock/rhythms
   affected by e-reading 70
   and body temperature 75
   and exposure to nature 56–8
   and flashing lights during sleep 82
   and tiredness at work 124
Citroën 109
citrus scent 20, 52, 124–5, 138, 159, 165–6
Classen, Constance: *Worlds of sense* 58
climate change 252
Clinton, Hillary 140–41
clock faces *8*, 9–10
clothing
   Christian Louboutin brand 238
   enclothed cognition 220
   high heels 237–8
   ladies dressing provocatively 226
   red underwear 219–20
   and sexual attraction 235–8
   sportswear and colour 214–20, 221, 266
   superhero attire 220, 221
   transmitting interpersonal tactile stimulation 257
Club Monaco 158
Coca-Cola 154
Cochrane Reviews 192, 200
coffee/caffeine 19, 67, 72, 78, 87, 127–8
   aromas 146–7
   coffee machines 36–7
   coffee shops 139n, 158
   and the 'coffice' 139n
   fragrance 20
   lavender oil and caffeinated mice 84
cognitive-behavioural therapy 68, 82

Cohen, Scott 260–61
cold pressor test 196
Colgate-Palmolive Company 87
colour 26–9
   with appetite-suppressing effects 32–4
   avoidance motivation and 'seeing red' 218–19
   and the brain 219
   of cars 99
   coloured light 33–4
   in hospitals 181, 184–5
   in kitchens 32–4
   red *see* red colour
   and sexual attraction 235–7
   and sleep 79
   smelling colour 148–50
   sportswear 216–20, 221, 266
   and taste of food 29, 34
   therapy 184, 198
   and thermal comfort 123
   and the workplace 125
commensality, and creativity 139–41
Conrad, C. 192
consumer engineering 10
consumer neuroscience 6, 10, 68, 142, 144
   and background music 150–54, 165–8
   and multisensory marketing *see* multisensory marketing
   and sense of smell 144–50, 165–7
   and subliminal seduction 154–6
cortisol, salivary 21
cosmetics
   'cosmetic neuroscience' 6
   and Covid-19 pandemic 3n
   global market 3n
   *see also* scents and perfumes
Covid-19 pandemic 2, 3n, 92, 257–9
Crawford, Ilse 25–6

creativity 126–7, 228–9
   and commensality 139–41
Crécy, Studio de 172
Crossmodal Research Laboratory, Oxford 103, 235, 243
Cyborg Nest, North Sense 260–61, *261*

*Daily Mail* 149
dance 225–6
Darwin, Charles 228–9
dating *see* sexual attraction and dating
David Lloyd gyms 203, 204
dawn chorus 57, 85
de Burgh, Chris (Christopher John Davison): 'The Lady in Red' 237
Decq, Odile 24–5
dementia patients 181–2, 183n, 195
dentists' surgeries
   drill sounds 190–91
   and scents 196–7
Department of Health Working Group on Arts and Health, UK 183
desktop pictures 135–8
*Dial M for Murder* 237
Dichter, Ernest 143
Dickens, Charles 90
Dietary Approaches to Stop Hypertension (DASH) 181
Disaronno 148–9
dishonest signalling 242–3
Djokovic, Novak 209
DNA, major histocompatibility complex 231, 244–5
Dobzhansky, Theodosius 8
Dodow 71
dopamine fasting 249–50
Dorsey, Jack 40–41, 90
Dove, Arthur 255
   *Fog Horns* 254
*Downton Abbey* 35
dress *see* clothing

Driver, Jon 187–8
drivers and travelling by car 92–115
  and car sickness 93, 112–14
  drowsiness and sleeping at the wheel 104–6
  effect of scenery and environment 106–9
  listening to music 99–100
  neuroscience and driver attention 103–4
  and reality of road conditions 92
  risk compensation 111–12
  road traffic accidents 102–3, 104, 108, 109
  and talking on the phone 100–102
  technology, safety and distraction 100–106
  texting at the wheel 102–4
driving games, video 96, 99
Dropbox 139
Dulux 3, 125
Dunkin' Donuts, 'Flavor Radio' campaign 147–8
Durex 3

earplugs 76, 79–80, 168–9, 265
  noise-cancelling headphones 130, 190
eating *see* food and eating
*Economist, The* 118–19, 233
Edmonds, Jayme 232
Elliot, A. J. 235, 236
enclothed cognition 220
energy drinks 213–14
Eno, Brian 193
essential oils 23n, 52, 84, 111, 212
eugenol 196
evolutionary psychology 21–2, 26n, 204, 219, 222, 225, 227, 229, 244–5
exercise and sport 199–221

'ancestral health' movement 203–4
  and body posture 217
  and the brain 199, 209–10, 214
  and clothing 214–20, 221, 266
  crowd noise 211–12, 218n, 266
  and energy drinks/mouthwashes 213–14, 221, 266
  frequency of taking exercise 199, 200
  grunting of tennis players 209–10, 266
  and gyms 201–3, 206–7
  indoor vs outdoor exercise 201–3
  and music 205–8, 266
  and nutritional supplements 213
  optimizing exercising environment 199–200, 202
  and smell 212
  smiling's effect on runners 213, 266
  and sonic feedback 208
  and well-being 200–201
  working-out with the senses 220–21
*Exxon Valdez* oil spill 88
eye contact 228, 250

Facebook 262n
facial attraction 227–8, 233, 234
  *see also* smiling
facial symmetry 227
family arguments 23
Fellowes, Future of Work report 120, *121*
Feng Shui 22
Ferrero Group 146
Field, Tiffany 1–2, 117, 194, 255
*Financial Times* 170
first night effect (FNE) 78–80
Fisher, M. F. K. 240
Fismer, K. L. 83
flowers 57–8

flowers – *cont'd.*
  and building odour 19, 20
  death, funerals and 20
  fragrance attracting higher spending in stores 159
  trade-off between looks and smell 58
  'well-being bouquets' 20
foetal loading 238
foetal soap syndrome 6
food and eating
  commensality and creativity in the workplace 139–41
  and effect of colour on taste 29, 34
  and effect of music on taste 29
  and effect of table settings on appetite and taste 34–5, 265
  effects of watching TV programmes whilst eating 35–6
  and gourmand fragrances 233
  in hospital 180–82
  seductive spicy meals 226–7
  before sleeping 67
football 215–16, 217–18
*Forbes* magazine 139
Ford, Henry 99
forest bathing 45–6
Fritz, T. H. 207–8
Froriep, Ludwig *248*

Galton, Francis 1
Gap 161
gardens 42–5, 53–4
  botanical 60–61
  hospital healing gardens 179
  and the nature effect 44–62
  Santandercito (author's garden retreat) 59–60, *59*
  walled 57–8
  'wild gardens' 60n
General Motors 95–6
geosmin 138

Gibbard, Roland 203
Glenmorangie whisky distillers 171–2
glucose mouthwash 213, 221, 266
glycogen 213
Goldsworthy, Dawn 19
Golub, Chris 151
Google 139
'Got milk' campaign 148
Grahn, P. 46
Great Depression 10
Greene, M. R. 62
Grindr 239
Griskevicius, V. 235
Guantánamo Bay *248*
*Guardian* 201
Gubbels, J. L.: *American highways and roadsides* 108
Guinness 169, 171
guitar playing 228, 246
gum 214
gyms 201–3, 206–7

habitat theory 21–2
Hadland, Hugh 94–5
Hamleys 159
happiness 47, 63, 228
Harbisson, Neil *259*, 260
Hardick, J. 104
Harlow, Harry 257n
*Harvard Business Review* 116
Harvey, Allison 69
healthcare 174–98
  animal therapy 195
  aromatherapy *see* aromatherapy
  and art 183–4
  auditory/sonic 188–93
  Cochrane Reviews 192, 200
  and exercise 200–201
  'experience economy' thinking 177–8
  and exposure to nature 51, 177–8 *see also* nature: effect

in hospital *see* hospitals
interpersonal touch 2, 193–5 *see also* massage
multisensory illusions in 185–8
multisensory medicine and sensory overload 197
music therapy 192–3
neuroscience and sleep-health industry 68
'odour therapy' and healing power of scents 195–7
private 177–8
quality of patient experience affecting treatment outcomes 177–8
and salutogenesis 179
Snoezelen concept 197–8
surgeons listening to music while operating 174–6
*see also* well-being
Healy, Mark 107
hearing
 beauty and the voice 242–3
 hearing what you see 14–15
 music *see* music
 nature's sounds *see* nature: sounds of
 noise *see* sound and noise
 overstimulation 1
 and psychoacousticians 95
 and superadditivity 15–16
 and white goods design 36–7
heliotropin 20
Heschong, Lisa: *Thermal delight in architecture* 160
Highways England 106
Hill, R. A. 219
Hirsch, Alan 159
Hollister Co. 158
Homburg, C. 167
home environment 18–41
 air quality 77, 120
 bathing *see* bathing
 bathrooms *see* bathrooms
 building odour 21
 ceiling height 23–4
 colours *see* colour
 curvilinear forms 22–3
 and family arguments 23
 gardens *see* gardens
 indoor plants 22, 76–7, 265
 indoor temperature 29–30, 76, 252
 interior design for multisensory mind 5–6, 21–41
 kitchens *see* kitchens
 lighting 33–4, 35
 music in 23, 29, 35
 open vs closed rooms 21–2
 and 'sensory living' 24–6
 and sleep *see* sleep
 tables *see* tables
 water consumption 41
hospitals 176–90, 191–3
 art in 183–4
 background noise 180, 189–90
 and colour 181, 184–5
 crockery colours 181
 dementia patients 181–2, 183n
 food and eating in 180–82
 healing gardens 179
 ICUs 183n, 189
 looking like high-end hotels 177–8, 183
 music and ambient soundscapes in 181–2, 192–3
 psychiatric 181–2
 and salutogenesis 179
 surgeons listening to music while operating 174–6
 VR distraction in 191–2
*Housing Space Standards* 37–8
Hug Shirt 257, *258*
HuggieBot 257

hugging 2, 257–8
Hugo Boss 159
Hultén, Bertil 162
Hyundai 100

ice-baths 196
IKEA stores 162
Imperial Festival, London 176
Inman, Dominic 181
insomnia 65, 68, 69, 76
  rebound 83
intelligent transportation systems 111
interior design 5–6, 21–41
ivy 76

James, William 14n
Jeanneret, Pierre 31
Jeffries, Mike 167
Jenner, Kendall 32
*Jezebel* 237
Jobs, Steve 9n
Johnson & Johnson 3
Johnson's baby powder 20
Jones, David 213
Jones, Natalie 140
*Journal of Retailing* 156
Joye, Y. 62
Juicy Couture 159
JWT Advertising Agency 253

Kandinsky, Wassily 255
Kaplan, Rachel 48–9
Kaplan, Stephen 48–9
Kenrick, D. T. 235
Kipchoge, Eliud 213
kitchens 30–31, 36–8
  appliances and noise in
    36–7
  and colour 32–4
  John Nash show kitchen, Brighton
    Pavilion 24

and solo living 38
table shapes 22–3
white goods design 36–7
Kneebone, Roger 174
Knoeferle, Klemens 151
Kotler, Philip 156, 165, 177, 254
Kramer, Art 204
Kreutz, G. 207

*Lancet* 228
*Landscape Research* 46
lap dancers 226
lavender 26, 52, 81, 82–4, 111, 124,
  165, 167, 197
Le Corbusier (Charles-Édouard
  Jeanneret) 21, 26–7, 30–31
Leder, Helmut 224–5
Lee, Ingrid Fetell 22
Lenochová P. 232
Levitin, Daniel: *The organized mind* 130
light/lighting 33–4, 35
  work 125
  blue light boosting alertness and
    cognitive performance 125
  dawn light and waking up 89–90
  light bulbs 33–4
  light hunger and SAD 118, 251
  sleep and blue/artificial light 69–71,
    82, 89
  sleep and flashing lights 82
  and tiredness/alertness at work 124
linalool 83–4
Lipton Ice drink 155
London Underground 148–9
Longoria, Eva 74
lordosis 238
lucid dreaming 84–5
lumbar curvature 238
Lush 159
Lynch, K. 108n
Lynx effect 232–4

M&M World store 146
MacBook Pro screens 162
machine learning 144
MacKay, Sir David 189–90
major histocompatibility complex 231, 244–5
Manchester United FC 215–16
Manhattan 118
masculinity 229–30, 234, 242
    see also testosterone
Maslin, Mark 30
massage 81, 117, 193–4
    aromatherapy 195, 255
    chairs 194
materials 25–6
*Matrix, The* 262–3
Masumoto, David Mas: *Four seasons in five senses* 53
*Maxim* 234
Mayo clinic 178, 183, 185
McCain's Ready Baked Jackets 149
McDonald's 144, 164
McGill Pain Questionnaire 191n
McGurk effect 15
Mehta, R. 126, 127
melatonin 69–70
menstrual cycle 226, 230–31, 242
Mercedes 109
mere exposure effect 6–7
Mieris, Frans van: *Lunch with Oysters and Wine* 240, 241
Milgram, Stanley 250
Milliman, Ronald E. 150, 151
Mindlab 35
mirror box/therapy 186–7, *187*
mobile phones see smart phones
Mood Media 151n
Moona pillow 76
Morrin, M. 165–6, 168
Moseley, Lorimer 186–7, 188
Mourinho, José 213

multiple messages hypothesis 244–5
multisensory attraction 3
multisensory illusions 185–8
multisensory judgements 209–10, 243
multisensory marketing 144, 146–54, 164–72, 252–3, 254–5
    and atmospherics see atmospherics in shopping
    with background music 150–54, 165–8, 266
    functionally subliminal cross-sensory marketing 155–6, *155*
    online marketing and shopping 170–72
    and smelling colour 148–50
    and in-store tasting 169
    and sub-additivity 16, 165–8, *166*
    with synaesthetic tasting 169, 171
multisensory mind, interior design for 5–6, 21–41
multisensory perception 11
    and merging of the senses 13–16
multisensory stimulation
    and babies 3
    balanced see sensory balance
    and exposure to nature 42–3
    and gyms 202
    and marketing see multisensory marketing
    and the nature effect 54–6, 251–2 see also nature: effect
    and our evolutionary niche 7–8
multitasking 1, 250
music 12, 23, 26, 35, 54–5
    background music at work 124, 266
    as courtship display 228–9
    drivers listening to 99–100
    elevator music 151
    with exercise and sport 205–8, 266
    in hospital 174–6, 181–2, 192–3

music – *cont'd.*
  and multisensory marketing 150–54, 165–8, 266
  'musical agency' 207–8
  in operating theatres 174–6
  and sexual arousal 224–5, 226
  and shopping 150–54, 165–8, 266
  and taste of food 29
  therapy 192–3
  for waking up 88
Musk, Elon 262
Muzak 151
Myer, J. R. 108n

Nadal, Rafael 209
Napoleon I 90
NASA 76–7
National Sleep Foundation 81
nature
  and attention restoration theory 48–9
  biodiversity 51–2
  and bodily rhythms 56–8 *see also* circadian clock/rhythms
  and the brain 49–50
  brought to the workplace 132–9
  deficit 46
  on desktop 135–8
  driving through 106–8
  effect 44–62, 251–2
  exercising in nature/virtual nature 202, 203–4
  feel of 53–4, 138
  forest bathing 45–6
  and the garden *see* gardens
  healthcare and exposure to 51, 177–8
  and multisensory stimulation 42–3
  ranking the benefits of 54–6
  and recovery from/reduction of stress 46, 132–3, 136n
  sensehacking nature for well-being 62–4
  sight of 50–51, 137–8
  simulations vs VR headsets 204–5
  smells of 52–3, 57–8, 138
  sounds of 51–2, 55, 57, 73, 74, 85, 130–31, 132, 265 *see also* birdsong
  synthetic/virtual reproductions of 52, 137–8
  taste of 53
*Nature* 66, 74, 75
Navratilova, Martina 210
near-rear peripersonal space 104
neural lace 262
Neuralink 262
neuroimaging 228
neuromarketing *see* consumer neuroscience
neuroscience 255
  affective 256
  cognitive 21, 112
  consumer *see* consumer neuroscience
  cosmetic 6
  and driver attention 103–4
  replication crisis 152
  and sleep-health industry 68
  superadditivity *see* superadditivity
*New York Times* 24–5, 107
Niesta, D. 236
Nightingale, Florence 179, 183–4, 188–9
Nike 215
noise *see* sound and noise
North, Adrian 151–2
North Sense 260, *261*
NozNoz *31*, 32
nutritional supplements 213

Ode (hunger-inducing scent-delivery system) 182
office design 118–19, 128–32, 138n

Oliva, A. 62
Olmsted, Frederick Law 45
*Only Fools and Horses* 35
operating theatres, music played in 174–6
Operation (game) 175–6
orchids, wild 58n
orienteering 42–3
Oscar Mayer bacon-scented wake-up app 86
Ottoson, J. 46
Oura Smart Ring 90
Oyer, J. 104
oysters 240–41

packaging, product 9, 143–4, 161, 163
Packard, Vance: *The hidden persuaders* 143
Padilla, José *248*
paint colours, interior decoration 26–9
Panera Bread 145
passive body heating (PBH) 75
Peck, Joanne 161
Pencavel, John 116
peppermint 52, 110, 124, 138, 199, 212, 220–21
perfumes *see* scents and perfumes
personal care products 41
Peugeot 308 GTi 96
phantom limb pain 185–7
Phelps, Michael 206
phenotypes 245
phytoncides 46
Pilkington, K. 83
Piqueras-Fiszman, Betina 163
Pixar 139
plastic trees 133
Plath, Sylvia 39
Pliny 58
pollution/pollutants 5, 77, 251

road traffic accidents and air pollution 109
sick building syndrome and air pollution 120
potted plants 22, 76–7, 132, 133–4, 183, 265
Power Energy toothpaste 87
predictive coding 214
Procter & Gamble 23n
product packaging 9, 143–4, 161, 163
prospect-refuge theory 21–2
psychoacousticians 95
Punto Blanco stores 157–8
Pzizz 74

RAND Corporation 66
Read, Lily 101
Reagan, Ronald 65
red colour
  avoidance motivation and 'seeing red' 218–19
  and the brain 219
  red light 28
  and sexual attraction 235–7
  sportswear 217–20
  underwear 219–20
redundant signals hypothesis 245
refuge theory 21–2
Renault 97
replication crisis 152
Ribas, Moon 259–60
Riseborough, Andrea 44n
risk compensation 111–12
road traffic accidents 102–3, 104, 108, 109
roads 106–8
Robbie, Margot 74
rocking 76
Roenneberg, Till 66
Romans 40n, 45
Rooney, Wayne 215–16

Roosevelt, Franklin D. 108
roses 58
Rusedski, Greg 209

salivary cortisol 21
Salsa, Fragrance Jeans 157
salutogenesis 179
Samsung Experience stores 159
San Pellegrino 155–6, *155*
Santandercito (author's garden retreat) 59–60, *59*
scented candles 19–20
scents and perfumes 12, 16, 18
   aromatherapy *see* aromatherapy
   body odour and scent of a woman 230–32
   bread smells 144–5, 266
   citrus 20, 52, 124–5, 138, 159, 165–6
   coffee aromas 146–7
   and dentists' surgeries 196–7
   and driving 108–11
   exposure to alerting ambient scent at work 124–5
   fabrics made to feel softer by scents 25–6
   of flowers and plants 20, 57–8
   fragrance industry 231
   gourmand fragrances 233
   hunger-inducing scent-delivery system, Ode 182
   lavender *see* lavender
   and 'new-car-smell' 93–5
   'odour therapy' and healing power of scents 195–7
   Oscar Mayer bacon-scented wake-up app 86
   peppermint *see* peppermint
   perfume displays in department stores 145n
   and personal care products 41
   pumped through ventilation systems 109, 140, 148, 159
   to reduce stress 124, 266
   scent-enabling technology 145–6
   scented baths 81
   scented candles 19–20
   and sexual attraction 243–4
   synthetic smells/scenting in stores 144–6, 157–8, 165–7
   and 'well-being bouquets' 20
   'youthful' fragrances 234–5
*Science* 51n, 87, 114, 126, 133
*Scientific American* 179
screensavers 135–8
Scriabin, Alexander 255
*Sea gulls . . . Music for rest and relaxation* 182
seasonal affective disorder (SAD) 6, 118, 251
Seidman, Dov 206
Selfridges 168
Senders, John W. 105–6
sensation transference 162–3
sensehacking
   and anthropomorphic design 8, 9–10
   consumer engineering as forerunner to 10
   creativity 126–7
   definition 4
   and driving *see* cars; drivers and travelling by car
   and exercise *see* exercise and sport
   future of 247, 259–64
   and the home *see* home environment
   and the marketplace *see* multisensory marketing; sensory marketing; shopping
   multisensory science of 16–17
   and nature *see* nature
   new sensations 259–62

present-day importance of 10
scope of use for well-being 4, 12–13 *see also* healthcare
and sensory crosstalk *see* sensory crosstalk/interaction/transference
and sexual attraction *see* sexual attraction and dating
and shopping *see* sensory marketing; shopping
simple sensehacks 265–6
and sleep *see* sleep
and the workplace *see* work and the workplace
sensism 253–7
sensory balance 1
  babies' need of 3
  lacking in workplace 118
  lacking with social isolation during Covid-19 pandemic 257–9
  and sensism 253–7
  *see also* multisensory stimulation
sensory crosstalk/interaction/transference 10–13, 165, 254–5
  and effect of colour on taste 29, 34
  effects of watching TV programmes whilst eating 35–6
  and exposure to nature 55–6
  hearing the quality 95–8
  hearing what you see 14–15
  and marketing *see* multisensory marketing
  music's effect on taste of food 29
  scents that make fabrics feel softer 25–6
  sensation transference 162–3
  and shopping atmospherics *see* atmospherics in shopping
  smelling colour 148–50
  table setting's effect on taste of food 34–5, 265
sensory deprivation 248–9
sensory dominance 14–15
sensory incongruence 16, 113, 165–8
sensory loss 2–3
  with forced deprivation 248–9
sensory marketing 142, 252–3
  and anthropomorphism 8, 9–10
  and atmospherics *see* atmospherics in shopping
  and interaction of senses 11 *see also* multisensory marketing
  multisensory *see* multisensory marketing
  and smell 144–50, 157–9, 165–7
  and superadditivity 15–16, 164–7
  white goods design and noise 36–7
sensory overload 1, 165–8, 197, 249–51
  and technology 1, 250
sensory processing disorders 250
sensory stimulation
  balanced *see* sensory balance
  below threshold of awareness 154–6
  and dopamine fasting 249–50
  multisensory *see* multisensory stimulation
  overload *see* sensory overload
  power used for well-being *see* sensehacking
  sleep problems correlating with poor diet of 251
  underload *see* sensory underload
  *see also specific senses*
sensory triggers 8–9, 20, 21, 135, 140, 171–2, 182, 218–19, 252, 253
sensory underload 2
  forced sensory deprivation 248–9
Seo, Han-Seok 33
sexual attraction
  and dating 222–46
  and 2D:4D digit ratio 230

sexual attraction – *cont'd*.
  aphrodisiacs 240–41
  arousal 223–7
  and 'beer goggles' effect 241–2
  and clothing 235–7
  and dance 225–6
  and dishonest signalling 242–3
  and evolutionary psychology 222, 225, 227, 229, 244–5
  and eye contact 228
  and facial symmetry 227
  and fertility indicators of female body 229, *229*
  and guitar playing 228, 246
  and lumbar curvature 238
  and masculinity 229–30, 234, 242
  and multiple messages hypothesis 244–5
  multisensory appeal 244–6
  and music 224–5, 226, 228–9
  online dating tips 239–40
  and oysters 240–41
  and smell 229, 230–32, 243–4
  and smiling 228, 246
  and spicy meals 226–7
  and thrillers 224, 266
  and the voice 229–30, 242–3
Shakira 225
Sharapova, Maria 209
*Shattered* 80–81
Sheldon, R. 98–9
Shimazu 139–40
shopping 142–73
  and atmospherics *see* atmospherics in shopping
  impulsive vs contemplative shoppers 168
  and music 150–54, 165–8, 266
  and neuromarketing 142
  online 142, 170–72
  and personalization 170

and sense of smell 144–50, 157–9, 165–7
and sensory marketing *see* multisensory marketing; sensory marketing
and in-store tasting 169
'til you drop 172–3
*see also* supermarkets
showering 38–9, 75, 265
sick building syndrome (SBS) 6, 119–20, 133–4, 251
sight 2–3
  and driving 105–6, 108n
  hearing what you see 14–15
  of nature 50–51, 137–8
  overstimulation 1
  and table setting's effect on taste of food 34–5, 265
Simba Sleep 89
Simpson, Peter 169
Singer, Isaac Bashevis 215
Sitwell, W. 34n
skin 2, 194, 195, 255–6
  galvanic skin response 28
  massage *see* massage
sleep 65–91
  and air temperature 76
  apps to help 68, 74, 89–90
  bathing/showering before 75, 81, 265
  and blue/artificial light 69–71, 82, 89
  and body temperature 75–6
  and colour 79
  cycles 67
  deprivation 65–6, 80–82, 90–91
  diaries 91
  drowsiness and sleeping at the wheel 104–6
  earnings correlated to weekly hours of 66–7
  engineering 67
  first night effect 78–80

food and drink before 67
health industry 68, 78
health problems linked with insufficient sleep 65–6, 82
hygiene 67, 90
inertia 86–8
insomnia *see* insomnia
and lavender scent 82–4
lucid dreaming 84–5
medications 68
neuroscience and sleep-health industry 68
nodding off and mental imagery 68–9
and noise 72–4, 76, 79–80
optimum hours 65
patterns ('larks' and 'night owls') 77–8
and pot plants 76–7
and problem-solving 74
problems correlating with poor sensory stimulation diet 251
and recuperation time 190n
research 65, 66, 69–71, 74, 75–6, 77–8, 82, 83–5, 87–9, 90–91
and rocking 76
and screen time 69–71
sense hacks for improving 67–85, 265
tracking 90
waking from 85–9
and warm feet 75–6
and well-being 65–7, 77–8, 81–2
*Sleep* 124
SleepBot 89
Sleepio 68
smart phones 69
  driving and talking on the phone 100–102
  texting at the wheel 102–4
  wake-up apps 85–6

smell, sense of
  and appetite suppressants 31–4
  babies turning towards aromas of foods mothers ate during pregnancy 7
  and the brain 80
  and building odour 18–20, 21
  and cooking odours 31–2
  and driving 108–11
  and exercise/sport 212
  and interpersonal touch 195
  loss of 3
  and 'new-car-smell' 93–5
  'odour therapy' 195–7
  and scents *see* scents and perfumes
  and sexual attraction 229, 230–32, 243–4
  shoppers influenced by 144–50, 157–9, 165–7
  smelling a person's age 231, 266
  smelling colour 148–50
  and the smells of nature 52–3, 57–8, 138
  and synthetic smells in stores 144–6, 165–7
smiling
  and attraction 228, 246
  effect on runners 213, 266
  use in product design 8, 9–10
smoke alarms 86
Snoezelen concept 197–8
social isolation 257–9
  with forced sensory deprivation 248–9
solo living 38
Somnox Sleep Robot 76
Somnuva 89
sonic feedback 208
Sonos 23
Sony Style store 159

Sony Style store – *cont'd.*
  ambient soundscapes in hospital 181–2, 192–3
  auditory/sonic aspects of healthcare 181–2, 188–93
  beauty and the voice 242–3
  brown noise 130–31
  of car doors and interior parts 97–8
  of car engines 95–7
  creativity and background noise 127
  crowd noise 211–12, 218n, 266
  of dentists' drills 190–91
  disturbing background noise in hospitals 180, 189–90
  grunting of tennis players 209–10, 266
  hearing *see* hearing
  hospitals' background noise 180
  music *see* music
  of nature 51–2, 55, 57, 73, 74, 85, 130–31, 132, 265 *see also* birdsong
  negative health consequences of exposure to noise 51
  noise-cancelling headphones 130, 190
  and open-plan offices 129–32
  pink noise 132
  and sleep 72–4, 76, 79–80
  sonic feedback 208
  from traffic 51, 55
  voice and sexual attraction 229–30, 242–3
  of white goods and appliances 36–7
  white noise 73, 76, 130n, 132
Southern, Clare 81
sport *see* exercise and sport
sportswear 214–20, 221, 266
Spotify 208
Stanton, Tasha 191
Starbucks 147
Stevenson, J. J. 30
*Streetcar Named Desire, A* 237

stress 46, 81
  brown noise for reducing 131
  and non-communicable diseases 117
  and open-plan offices 128
  'positive stress' 40–41
  recovery/reduction through exposure to nature 46, 132–3, 136n
  and salivary cortisol 21
  scents to reduce 124, 266
  signs when driving 111
  in traffic jams 92
  Trier Social Stress Test 21, 137–8
  work-related 117, 128
sub-additivity 16, 165–8, *166*
subliminal messaging 154–6
Subway 145
suicide 3
sundowning 183n
superadditivity 15–16, 164–7
Superman 220
supermarkets 144–5, 150, 151–3, 155–6, 169, 257
synaesthesia 19n
  synaesthetic tasting experiment 169, 171

tablecloths 34–5, 265
tables
  effect of settings on taste of food 34–5, 265
  shapes of kitchen/dining tables 22–3
Taconic State Parkway Commission 107
tactile contamination 163–4
Takasago 234–5
Tanizaki, Jun'ichirō 24
  *In praise of shadows* 60
Taoists 45
taste, sense of
  effect of colour on 29, 34
  music and taste of food 29

and nature's taste 53
and in-store tasting 169
table setting and taste of food 34–5, 265
technology
    alerts for drowsy drivers 104–6, 110
    car 92–3
    and multitasking 1, 250
    and safe/dangerous driving 100–106
    scent-enabling 145–6
    screensavers 135–8
    and sensory overload 1, 250
    virtual reality headsets 191, 204
temazepam 68
temperature
    of bath water before sleep 265
    body temperature 75–6
    in the home 29–30, 76, 252
    in shops 160
    for sleep 76
    in the workplace 121–3, 265
tennis players, grunting 209–10, 266
Tesla 115
testosterone 219, 230, 241, 242
texting, at the wheel 102–4
textures 25–6
Thatcher, Margaret 65
thermal comforting, in the workplace 121–3
Thomas Pink stores 157
Thorntons 146
Three Mile Island nuclear disaster 88
Tinder 239
toothpaste 72, 87
touch
    animal therapy 195
    and Covid-19 pandemic 2
    elderly people not being touched 2
    feel of cars 98–9
    feel of nature 53–4, 138
    hugging 2, 257–8

'hunger' 1–2, 193, 255
interpersonal 2, 193–5 *see also* massage
and shopping 161–4
and smell 195
tactile contamination 163–4
and textures 25–6
Tracey, Irene 191–2
transhumanism 262n
Traum, Thomas 172
Treib, Marc 43
Treisman, Michel 113–14
Trier Social Stress Test (TSST) 21, 137–8
Tripp, Peter 80–81

UK Biobank 77
Ulrich, Roger 47–8, 51n, 136n, 179
Unilever 3
    Lynx effect 232–4
Uniqlo 147, 158
Updike, John: *Pigeon feathers and other stories* 238
Ursitti, Clara 243

vanilla 19
ventriloquism effect 14
Venus of Willendorf 229, *229*
VF Corporation 3
Vicary, James 154
Victoria's Secret 159
virtual reality 262–3
    headsets 191, 204
vision *see* sight
volatile organic compounds (VOCs) 119, 133–4
Volkswagen Golf 96, 97
VR *see* virtual reality

waking up 85–9
Walker, Matthew: *Why we sleep* 66
*Wall Street Journal* 145
Wansink, Brian 140
washing-up liquids 23n
watches *8*, 9–10
water consumption 41
Wedgwood, Josiah 118
Weitz, Julie 172
well-being 3, 62–4
  and art 183–4
  at work 118, 128–39
  and being touched 2
  and colour 28
  and exercise 200–201
  and experiencing nature 44–64
  healthcare *see* healthcare
  in hospital *see* hospitals
  and interpersonal touch 2, 193–5
  and scents 18, 20
  sensory stimulation used for *see* sensehacking
  and sleep 65–7, 77–8
  'well-being bouquets' 20
  *see also* happiness
WeWork 127
white coat hypertension 183
white goods design 36–7
*Who Framed Roger Rabbit?* 237
Wilder, Gene: *The Woman in Red* 236
Williams, Florence: *The nature fix* 204
Williams, Serena 209
Williams, Venus 209
Wilson, W. O., biophilia hypothesis 44–5, 47, 49, 52, 60–61
wine 151–3

work and the workplace 116–41
  air-conditioning/temperature and sexism/gender 121–3, 265
  background music at 124, 266
  bringing nature to 132–9
  carbon dioxide levels 120, 134
  and colour 125
  commensality and creativity 139–41
  and disengagement 117
  lighting 125
  meetings 126–8
  office design 118–19, 128–32, 138n
  open-plan workspace 128–32
  and scents 124–5
  and sensehacking creativity 126–7, 139–41
  and sensory imbalance 118
  and sick building syndrome 6, 119–20, 133–4, 251
  and stress 117, 128
  tiredness at 123–5
  well-being at 118, 128–39
  windows 137
  working hours 116–17
World Green Building Council 134
World Health Organization 72–3, 119, 189

Yahoo 139
Youssef, Jozef 174
YouTube 74
Yumetai 33

Zara 159
Zeez 89
Zellner, Debra 153
Zhu, R. 126